Chemistry of Bipyrazoles: Synthesis and Applications

Authored by

Kamal M. Dawood

&

Ashraf A. Abbas

Department of Chemistry, Faculty of Science,
Cairo University
Giza 12613, Egypt

Chemistry of Bipyrazoles: Synthesis and Applications

Authors: Kamal M. Dawood and Ashraf A. Abbas

ISBN (Online): 978-981-5051-75-9

ISBN (Print): 978-981-5051-76-6

ISBN (Paperback): 978-981-5051-77-3

need for a court order if at any point you breach any terms of this License Agreement. In no event will any delay or failure by Bentham Science Publishers in enforcing your compliance with this License Agreement constitute a waiver of any of its rights.

3. You acknowledge that you have read this License Agreement, and agree to be bound by its terms and conditions. To the extent that any other terms and conditions presented on any website of Bentham Science Publishers conflict with, or are inconsistent with, the terms and conditions set out in this License Agreement, you acknowledge that the terms and conditions set out in this License Agreement shall prevail.

Bentham Science Publishers Pte. Ltd.
80 Robinson Road #02-00
Singapore 068898
Singapore
Email: subscriptions@benthamscience.net

**BENTHAM
SCIENCE**

CONTENTS

PREFACE

Pyrazole is one of the most valuable nitrogen-based heterocycles and is incorporated in the constitution of a wide range of pharmaceuticals and agrochemicals. Direct connection of two pyrazole units produces six different bipyrazole skeletons that can be classified as i) *N-N* bond connected 1,1`-bipyrazoles; ii) *C-N* bond connected 1,3`- and 1,4`-bipyrazoles and iii) *C-C* bond connected 3,3`-, 3,4`- and 4,4`-bipyrazoles.

This book presents the recent achievements in the synthetic platforms toward the directly connected bipyrazole systems and their applications in academic, industrial, and material science fields. The construction of the targeted bipyrazole heterocycles was carried out *via* a wide-range of synthetic routes that grasp the attention of graduate and postgraduate chemists and pharmacists and material science researchers to make more efforts in this area to reach high impact findings for their applications in our life.

Most of the reported bipyrazoles are highly bioactive heterocycles demonstrating a broad array of significant inhibitory activities against several human diseases and agricultural pesticides and herbicides. They also have considerable applications in the material science area *via* involvement in the construction of metal-organic frameworks (MOFs) with distinguished industrial applications.

This book is presented in five chapters describing the synthesis of six connected bipyrazole systems and their brilliant and vibrant applications. As a result, we expect that the provided book chapters will be of pronounced support and a valuable source for the scientific community for developing new bipyrazole-based fascinating candidates towards optimization of their pharmacological benefits in the treatment of diseases as well as building up new MOFs for daily life applications that serve the humanity and industry.

We hope that the researchers and readers will find new ideas based on the provided work. Finally, we are very thankful to the Bentham Science Publishers for giving us the chance to publish this book.

CONSENT FOR PUBLICATION

Not applicable.

CONFLICT OF INTEREST

The authors declare that there is no conflict of interest.

ACKNOWLEDGEMENT

Declared none.

Kamal M. Dawood
&
Ashraf A. Abbas

Department of Chemistry, Faculty of Science,
Cairo University
Giza 12613, Egypt

CHAPTER 1

Chemistry of *N,N*- and *C,N*-Linked Bipyrazole Derivatives

Abstract: The synthetic routes to three differently connected bipyrazole systems, namely; 1,1`-, 1,3`- and 1,4`-bipyrazoles were reported. The main synthetic platforms were cyclocondensation reactions. Many of the reported bipyrazole derivatives had potent applications in material science as well as in pharmaceutical fields.

Keywords: 1,1`-bipyrazoles, 1,3`-bipyrazoles, 1,4`-bipyrazoles, Cross-coupling, Cyclocondensation, Nitrilimines.

1. INTRODUCTION

Bipyrazoles are nitrogen heterocycles that are consisted of two pyrazole moieties connected directly by a covalent sigma bond without any space linker. In this chapter, the considered connections are either *N,N*- or *C,N*-connection types. The *N,N*-linked bipyrazoles are named as 1,1`-bipyrazoles, and those *C,N*-bonded compounds are named as either 1,3`-bipyrazoles or 1,4`-bipyrazoles as shown in Scheme (**1**).

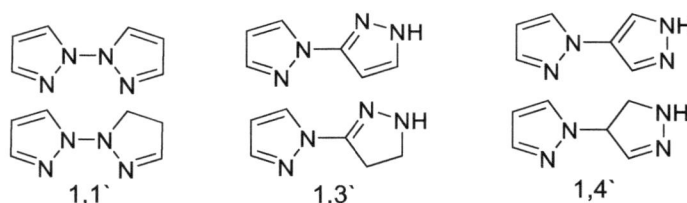

Scheme (1). The directly connected *N,N*- and *C,N*- bipyrazole systems.

The fulfilling pathways are: 1) reactions of tetracarbonyl or dihydroxydicarbonyl building units with hydrazines, 2) reaction of pyrazoles having a difunctional-side arm with hydrazines, 3) reaction of pyrazolyl-hydrazines with difunctional compounds (*e.g.* dicarbonyl, hydroxycarbonyl, ketonitrile or dinitrile substrates), and 4) metal catalyzed C-C cross coupling reactions of pyrazoles *via* C-H activation (Scheme **2**).

Kamal M. Dawood and Ashraf A. Abbas

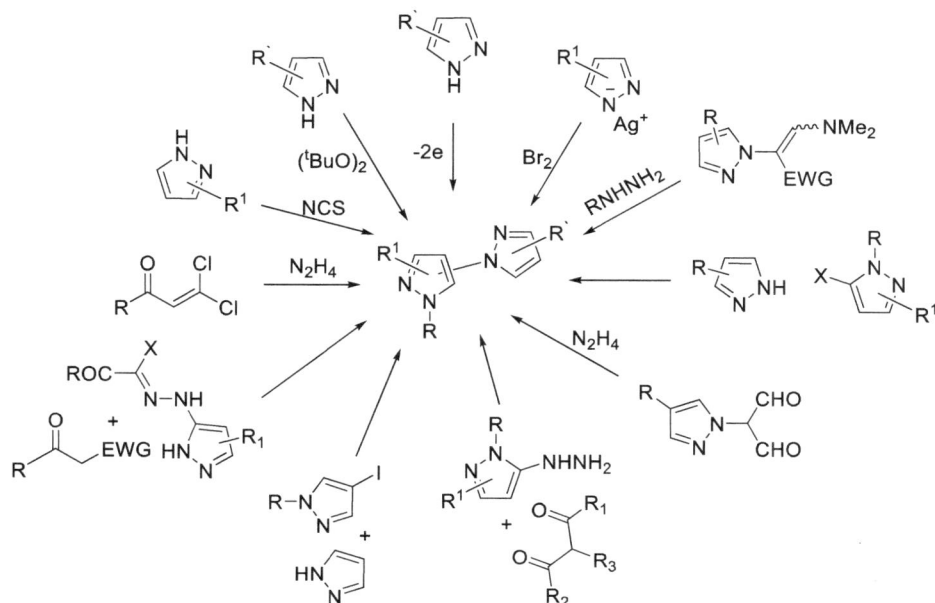

Scheme (2). The possible synthetic routes to *N,N*- and *C,N*-bipyrazoles.

Pyrazoles are one of the most abundant nitrogen heterocyclic compounds that have huge pharmaceutical and agro-chemical industrial applications [1 - 6]. Bipyrazoles are also a very interesting bioactive class of heterocycles that had pronounced biological activities. Particularly, 1,3`-bipyrazole derivatives had potent inhibitory activities against various diseases. For example, they exhibited cytotoxic [7], antimicrobial [8], anti-inflammatory [9] and antidiabetic activities [10] as well as herbicidal activities with excellent weed-controlling effects [11 - 13], potential agricultural pesticides [14, 15]. On the other hand, several 1,4`-bipyrazole derivatives were reported to have pronounced cytotoxicity activities [16] and for the treatment of Parkinson's disease [17]. The 1,4`-bipyrazole derivatives were employed as efficient ligands in the palladium-catalyzed C-N and C-O cross-coupling reactions of aryl halides with urea and with primary alcohols derivatives [18 - 22].

2. SYNTHESIS OF BIPYRAZOLE SYSTEMS

2.1. Synthesis of 1,1`-Bipyrazoles

Formation of the 1,1'-bipyrazole derivative **2** was performed by photolysis of ethyl 5-amino-3-(phenylamino)pyrazole-4-carboxylate **1** with *tert*-butyl peroxide or with dibenzoyl peroxide under mild reaction conditions. The reaction took place *via* radical dimerization of the pyrazole **1** (Scheme 3) [23].

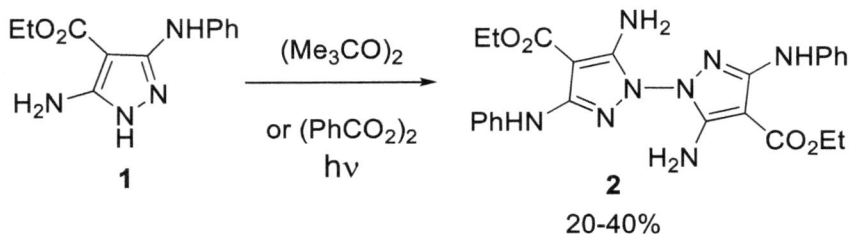

Scheme (3). Synthesis of 1,1`-bipyrazole **2**.

The dihydro-1,1'-bipyrazole derivative **6** was obtained from the reaction of 3-methoxycarbonyl-2-pyrazoline **3** with lead tetraacetate in benzene at 60°C. The reaction proceeded *via* the pyrazoline intermediate **4** which underwent further attack on **3** to give **6** in 17% yield. The ^{13}C NMR of compound **6** showed five peaks δ 52.3, 109.1 129.4 142.3 161.3 ppm due to OCH$_3$, pyrazole-carbons (C-4, C-5 and C-3) and C=O, respectively. The oxidation of **6** with *N*-bromosuccinimide (NBS) in refluxing carbon tetrachloride in the presence of a few drops of dry pyridine resulted in the formation of the symmetrical 1,1`-bipyrazole **7** in 55% yield (Scheme 4) [24].

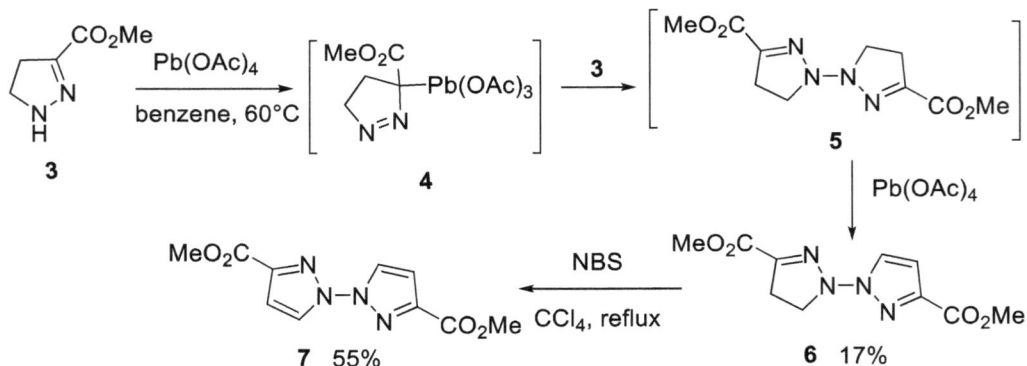

Scheme (4). Synthesis of 1,1`-bipyrazole **7**.

2.2. Synthesis of 1,3`-bipyrazoles

The 1,3-bipyrazole derivative derivatives **10** were synthesized, in good yields, from the reaction of the hydrazino-pyrazole derivative **8** with various symmetrical and unsymmetrical 1,3-dicarbonyl compounds **9** in the presence of 5% HCl (Scheme 5) [7]. The ^{1}H NMR spectrum of compound **10** (R$_1$=R$_2$= Me, R$_3$= H) displayed five singlet peaks at δ 2.16, 2.61, 3.32, 3.67 (due to four CH$_3$ protons) and 6.11 due to CH-proton and its ^{13}C NMR exhibited nine peaks at δ 11.0

$(C_5\text{-}CH_3)$, 13.7 $(C_{5'}\text{-}CH_3)$, 14.4 $(C_3\text{-}CH_3)$, 36.5 $(N\text{-}CH_3)$, 107.5 $(C_4\text{-}H)$, 134.9 (C_5), 143.6 $(C_{5'})$, 145.9 (C_3), 152.7 $(C_{3'})$ ppm.

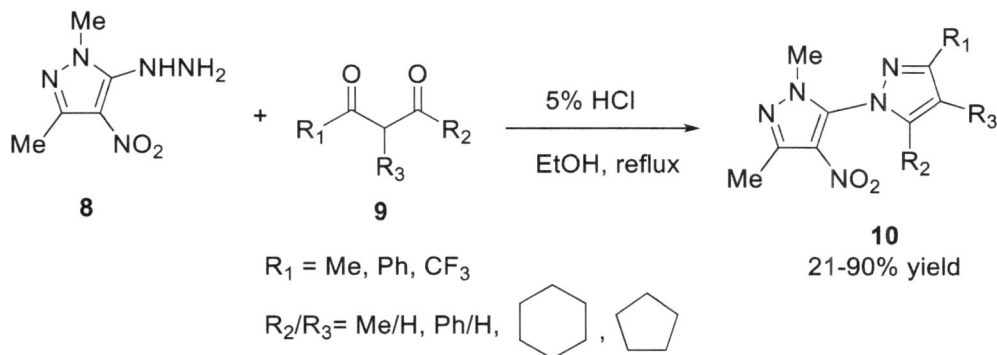

Scheme (5). Synthesis of 1,3`-bipyrazoles **10**.

Reaction of 3-hydrazino-5-methylpyrazole **11** with acetylacetone **12** furnished 3,5,5'-trimethyl-1'*H*-1,3'-bipyrazole **13** in high yield. Methylation of **13** with methyl iodide in the presence of *t*-BuOK led to the formation of 1',3,5,5'-tetramethyl-1'H-1,3'-bipyrazole **14** in high yield (Scheme **6**) [25 - 27].

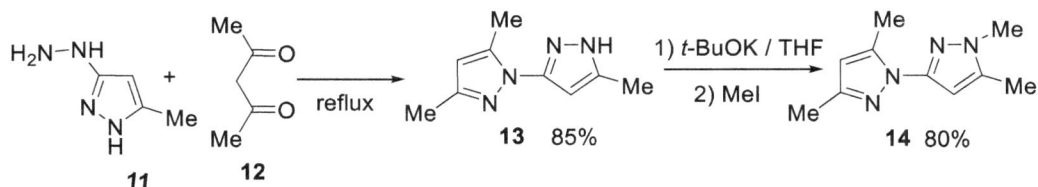

Scheme (6). Synthesis of 1,3`-bipyrazoles **14**.

In addition, when 3-hydrazinopyrazole **15** was treated with the benzoylpyruvate ester **16**, it afforded the 1,3'-bipyrazole ester derivative **17** in 36% yield. Methylation of **17** was carried out in the presence of *t*-BuOK to give the corresponding methylated 1,3'-bipyrazole derivative **18** in 29-61%. Treatment of **18** with lithium aluminium hydride in THF afforded the corresponding 3-hydroxymethyl-1,3'-bipyrazole derivative **19** in excellent yield (Scheme **7**) [28, 29].

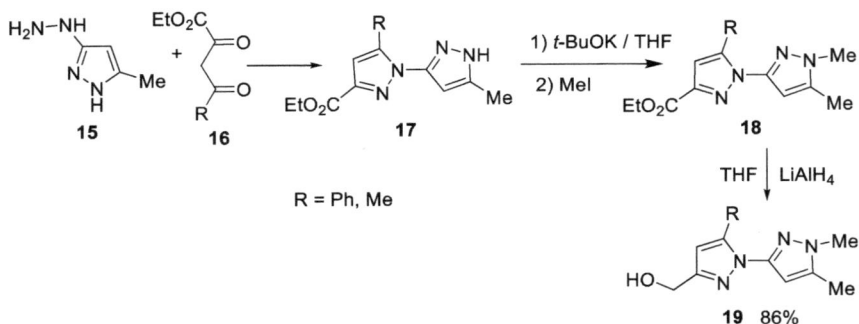

Scheme (7). Synthesis of 1,3`-bipyrazoles **19**.

The 1,3`-bipyrazole-based macrocycle **25** was synthesized according to the synthetic route described in Scheme (**8**). Thus, the reaction of ethyl 1,3`-bipyrazole-2-carboxylate **17** with 1,3-dibromopropane (**20**) in the presence of *t*BuOK as a base gave the *bis*-bipyrazole product **21,** which up on reduction with lithium aluminium hydride afforded the *bis*-hydroxymethyl-*bis*-bipyrazole product **22**. The latter *bis*-diol **22** was converted into the *bis*-chlorinated derivative **23** when treated with thionyl chloride. Condensation of **23** with 2-(4-aminophenyl)ethanol **24** resulted in the formation of the macrocycle **25** in 54% yield (Scheme **8**) [30].

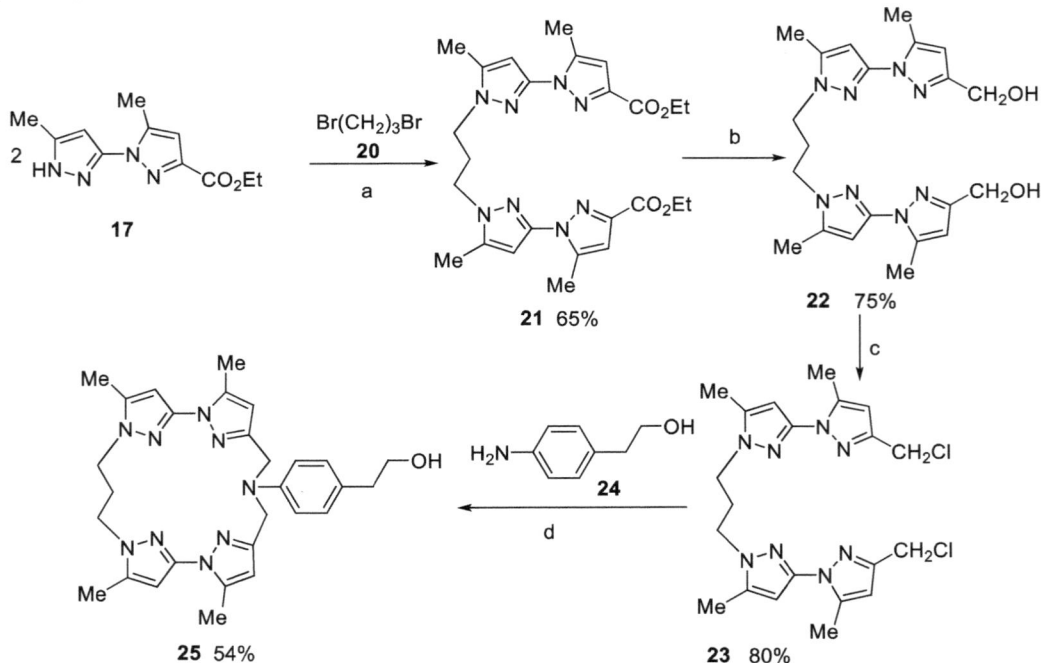

Reagents and conditions (a) *t*BuOK, THF, Reflux, 15 h, (b) LiAlH$_4$, THF, Reflux, 6 h, (c) SOCl$_2$, CH$_2$Cl$_2$, RT, 12 h, (d) Na$_2$CO$_3$; CH$_3$CN, Reflux, 24 h.

Scheme (8). Synthesis of 1,3`-bipyrazole-based macrocycle **25**.

The 1,3'-bipyrazole derivatives **29** were prepared by cyclocondensation of 5-hydrazinopyrazole derivative **26** with 1,3-dicarbonyl compounds **27** and **28**. Electrophilic substitution reactions of the latter 1,3'-bipyrazole **29** (R=R^1= Me) took place at position-4 of the pyrazole ring to give the corresponding 1,3'-bipyrazole derivatives **30** in high yields. Condensation of the pyrazol-5-ylhydrazine **26** with ethyl 2-cyano-3-ethoxyacrylate **31** afforded the 1,3'-bipyrazole derivative **32** in 69% yield (Scheme **9**) [31]. The ^{1}H NMR of **32** showed the following peaks: δ 1.34 (t, 3H, J = 7 Hz, CH$_2$CH_3), 2.68 (s, CH$_3$), 4.28 (q, 2H, J = 7 Hz, CO$_2$CH_2CH$_3$), 5.38 (s, 2H, NH$_2$), 7.33 (s, 5H, ArH`s) and 7.70 (s, 1H, pyrazole H-3) ppm.

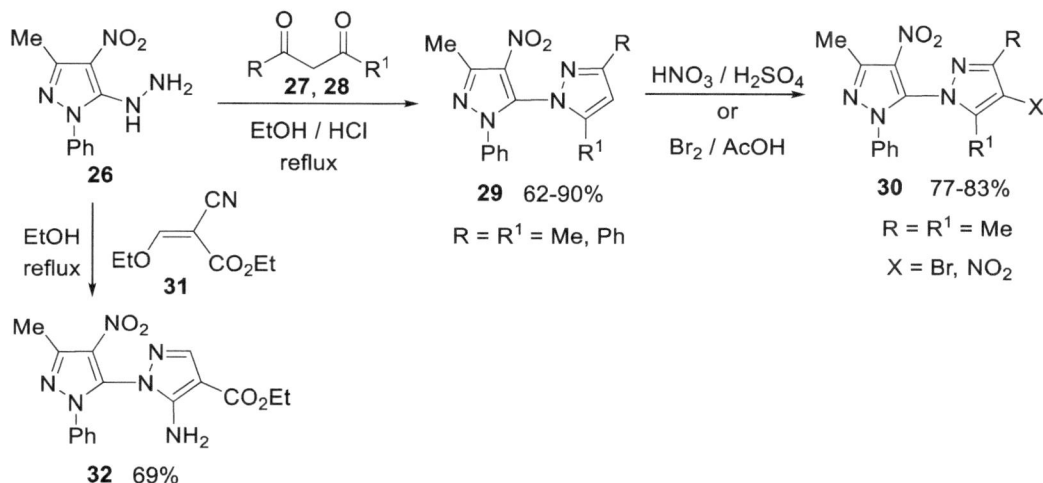

Scheme (9). Synthesis of 1,3'-bipyrazoles **29, 30** and **32**.

Heating the 5-hydrazino-1,3-oxazole-4-carbonitrile derivatives **33** with acetylacetone **34** led to the formation of the corresponding 5-(pyrazol-1-yl)-1,3-oxazole derivatives **35** in good yields. When the latter 5-(pyrazol-1-yl)-1,3-oxazole derivatives **35** were treated with hydrazine hydrate in refluxing ethanol, it afforded the corresponding 1,3′-bipyrazoles **37** in moderate to high yields. Formation of the 1,3′-bipyrazoles **37** took place *via* the ring opening of 1,3-oxazole of **35** to give the intermediates **36,** which underwent intramolecular cyclization to give **37** as depicted in Scheme (**10**) [32, 33].

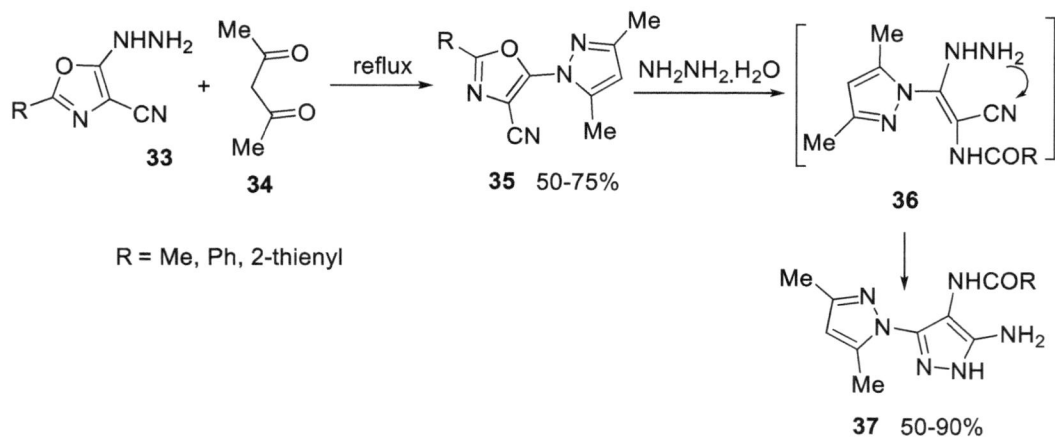

Scheme (10). Synthesis of 1,3`-bipyrazoles **37**.

Treatment of the hydrazino-pyrazole derivatives **38** with ethoxymalononitrile (**39**) in DMF afforded 1,3`-bipyrazole-adducts **42** in 87-91% yield. Formation of the bipyrazole derivatives **42** from the reaction of **38** with ethoxymalononitrile (**39**) was assumed to proceed *via* an initial addition of the amino group in pyrazoles **38** to the olefinic moiety in **39** to give the corresponding acyclic intermediates **40**, which underwent an intramolecular cyclization by loss of ethanol and aromatization to give the final products **42** (Scheme **11**) [34].

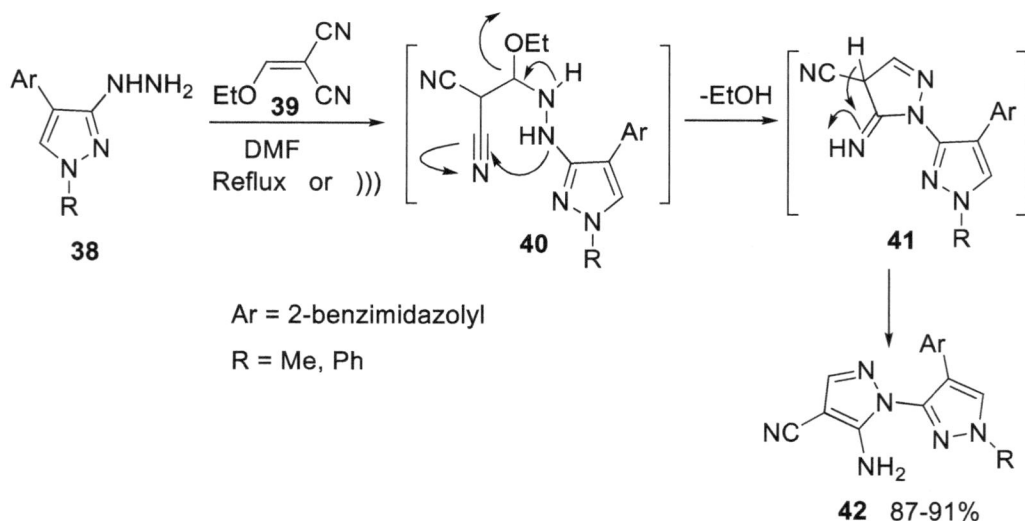

Scheme (11). Synthesis of 1,3`-bipyrazole derivatives **42**.

The 1,3'-bipyrazole derivatives **46** were prepared in good yields by cyclocondensation of the acrylonitrile derivatives **44** with 3-pyrazolylhydrazines **43** in the presence of potassium carbonate in refluxing ethanol (Scheme **12**) [35, 36]. Formation of the bipyrazoles **46** was supposed to take place *via* the intramolecular cyclization of the intermediates **45**.

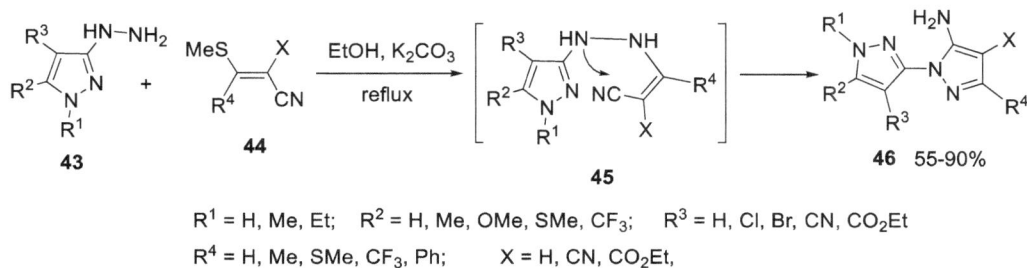

R^1 = H, Me, Et; R^2 = H, Me, OMe, SMe, CF$_3$; R^3 = H, Cl, Br, CN, CO$_2$Et

R^4 = H, Me, SMe, CF$_3$, Ph; X = H, CN, CO$_2$Et,

Scheme (12). Synthesis of 1,3`-bipyrazole derivatives **46**.

Bromination of a cold solution of the silver salt of pyrazole **47** in ether at 0 °C resulted in the formation of the 1,3`-bipyrazole derivative **48** as outlined in Scheme (**13**) [37].

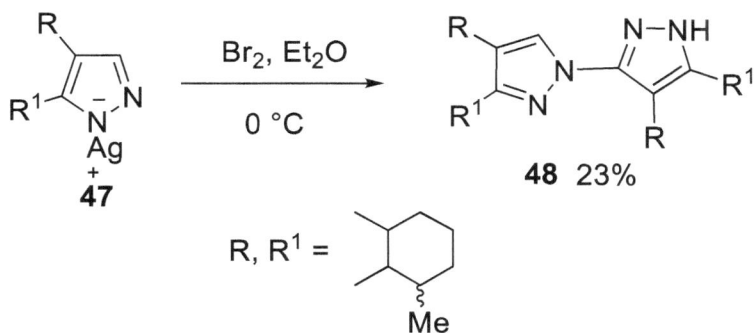

Scheme (13). Synthesis of 1,3`-bipyrazole derivatives **48**.

Electrochemical chlorination of pyrazole **49** in aqueous sodium chloride solution in the presence of CHCl$_3$ on platinum anode at a current of 3 A and 15 °C resulted in the formation of 4-chloropyrazole **50,** which upon dimerization under the reaction condition furnished 4,4'-dichloro-1,3'-bipyrazole **52** *via* the non-isolable intermediate **51** (Scheme **14**) [38].

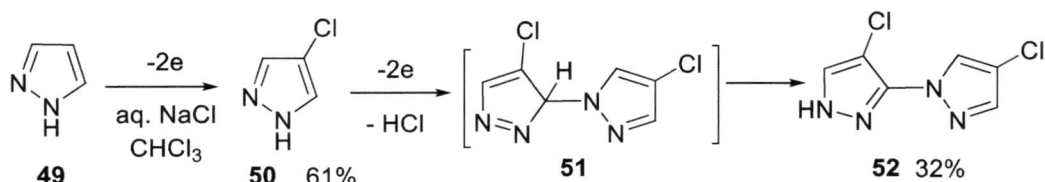

Scheme (14). Synthesis of 1,3`-bipyrazole derivative **52**.

Treatment of ethyl 3-ethoxypyrazole-4-carboxylate **53** with *N*-chlorosuccinimide (NCS) under microwave irradiation at 130 °C in dichloroethane furnished the 1,3'-bipyrazole derivative **55** in moderate yield. Occurrence of **55** was supposed to proceed *via* the hydrolysis of the chlorinated ethoxypyrazole moiety of the intermediate **54** (Scheme **15**) [39, 40].

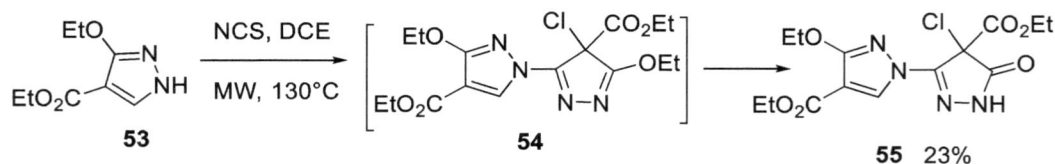

Scheme (15). Synthesis of 1,3`-bipyrazole derivative **55**.

Nucleophilic substitution of the activated 5-chloropyrazoles **56** underwent at its chlorine atom with pyrazole **57** (as a nucleophile) in dimethylsulfoxide (DMSO) at room temperature and gave the 1,3'-bipyrazole derivatives **58** in good yields. Reduction of the nitro group in compounds **58** was performed using NaBH$_4$/SnCl$_2$ followed by treatment with methanesulfonyl chloride and pyridine in dichloromethane to afford the 4'-(methylsulfonylamino)-1,3'-bipyrazole derivatives **59** (Scheme **16**) [41, 42].

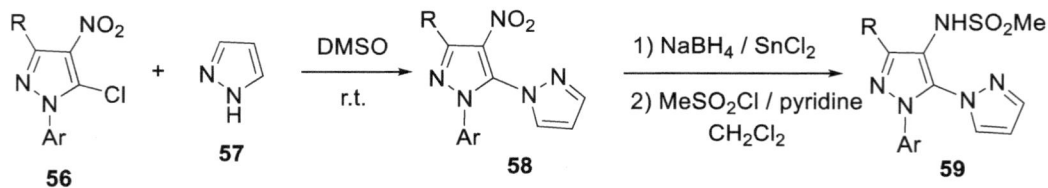

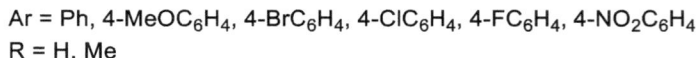

Ar = Ph, 4-MeOC$_6$H$_4$, 4-BrC$_6$H$_4$, 4-ClC$_6$H$_4$, 4-FC$_6$H$_4$, 4-NO$_2$C$_6$H$_4$
R = H, Me

Scheme (16). Synthesis of 1,3`-bipyrazole derivatives **59**.

Treatment of pyrazole **57** with 1,4-dinitropyrazole **60** in acetonitrile at ambient temperature led to the formation of 4'-nitro-1,3'-bipyrazole **61** in excellent yield.

The reaction occurred through a *cine*-substitution reaction where the entering group (pyrazole **57**) occupied position-2 adjacent to the leaving group (NO_2). Heating **61** with nitric acid in a mixture of acetic acid and acetic anhydride at reflux led to the formation of 1',4',4-trinitro-1,3'-bipyrazole **62** in high yield (Scheme **17**) [43, 44].

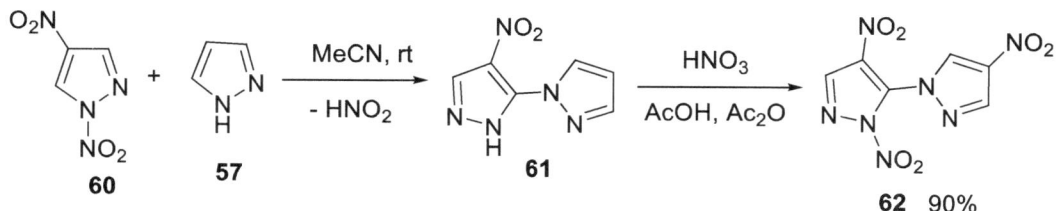

Scheme (17).   Synthesis of 1,3'-bipyrazole derivative **62**.

Nucleophilic substitution reaction of 1-methyl-3,4,5-trinitropyrazole **63** with 1*H*-pyrazole derivatives **57a-c** proceeded regiospecifically at the 5-position. The reaction was conducted at room temperature in the presence of NaOH to afford the corresponding 1,3'-bipyrazole derivatives **64** in high yields (Scheme **18**) [45].

Scheme (18).   Synthesis of 1,3'-bipyrazole derivatives **64**.

When 2,6-dimethyl-1-(2-methylpyrazol-1-yl)-4-phenylpyridinium bis-tetrafluoroborate **65** was treated with pyrazole **57** in water at room temperature, it afforded 1'-methyl-1,3'-bipyrazole **68** in 73% yield. Formation of **68** took place *via* loss of pyridinium tetrafluoroborate **67** from the intermediate **66** under the reaction conditions (Scheme **19**) [46]. The ^{1}H NMR spectrum of compound **68** in $CDCl_3$ showed the following peaks: δ 3.80 (s, 3H, CH_3), 6.21 (1H, $J = 2$ Hz, H_4), 6.41 (1H, H_4'), 7.58 ($J = 2.5$ Hz, H_5'), 7.43 (1H, H_3), 7.71 (1H, $J = 1.8$ Hz, H_3'), and its ^{13}C NMR showed seven peaks corresponding to the methyl group and the bipyrazole-carbons at 30.7, 100.1, 107.2, 131.3, 138.1, 139.0 and 141.9 ppm.

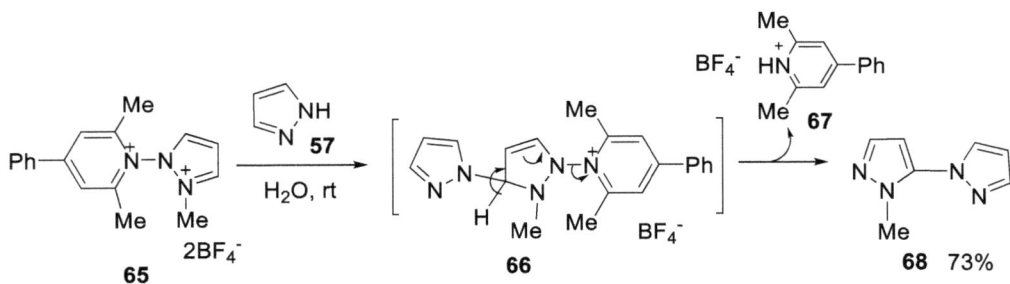

Scheme (19). Synthesis of 1,3`-bipyrazole derivative **64**.

Microwave irradiation of a mixture of 1-*tert*-butyl-5-chloro-1*H*-pyrazole-4-carboxylic acid-*N*-(2-adamantyl)amide **69** with pyrazole **57** using KF as a base and DMSO solvent led to the formation of 1'-*tert*-Butyl-1,3'-bipyrazoly--4'-carboxylic acid-*N*-(adamantan-2-yl)amide **70** (Scheme **20**) [47].

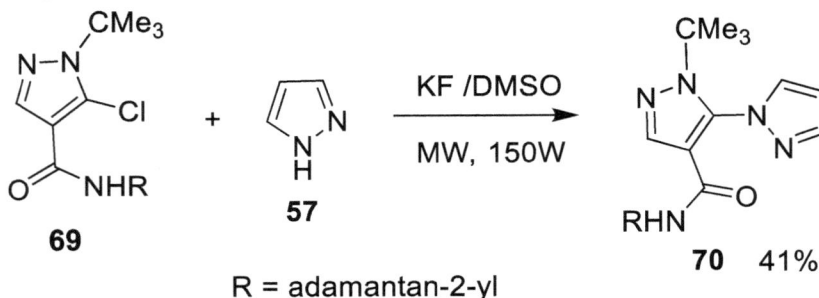

R = adamantan-2-yl

Scheme (20). Synthesis of 1,3`-bipyrazole derivative **70**.

Reaction of 3,3-dichlorovinyl methyl ketone **71** with hydrazine afforded the 1,3'-bipyrazole derivative **73** in reasonable yield. The reaction proceeded *via* the formation of the intermediate 5-chloro-3-methylpyrazole **72** followed by dimerization under the basic reaction condition with loss of HCl (Scheme **21**) [48].

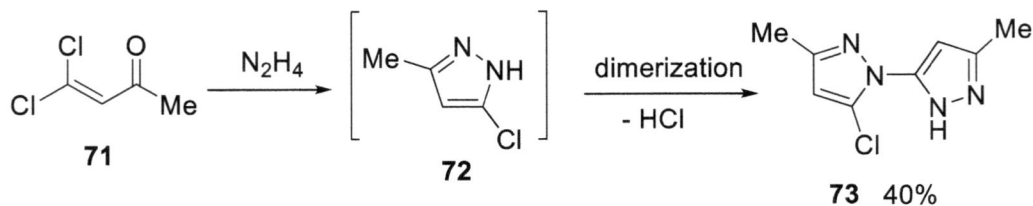

Scheme (21). Synthesis of 1,3`-bipyrazole derivative **73**.

The 4-bromo-3-phenylpyrazol-5-ylhydrazonyl chloride **74** was reported to react with the active methylene compounds **75** and **76** in ethanolic sodium ethoxide solution at room temperature to give the 1,3`-bipyrazole derivatives **77** and **78**, respectively (Scheme **22**) [49 - 51].

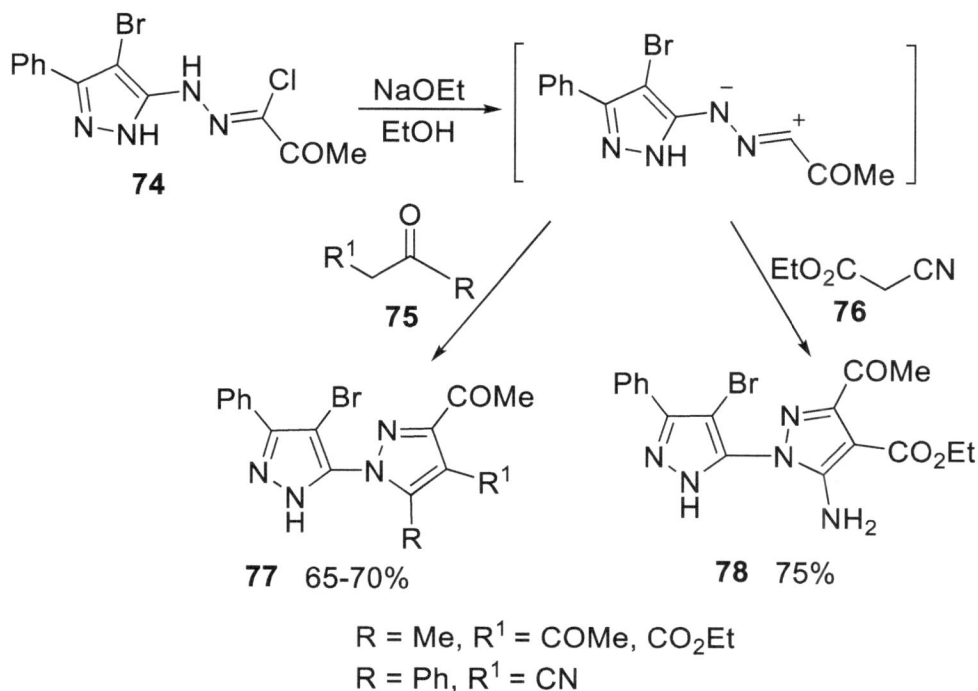

$$R = Me, R^1 = COMe, CO_2Et$$
$$R = Ph, R^1 = CN$$

Scheme (22). Synthesis of 1,3`-bipyrazole derivatives **77** and **78**.

2.3. Synthesis of 1,4`-bipyrazoles

Condensation reaction of 2-(4-nitro-1*H*-pyrazol-1-yl)malonaldehyde **79** with hydrazine hydrate led to the formation of 4-nitro-1'*H*-1,4'-bipyrazole **80** (Scheme **23**) [52].

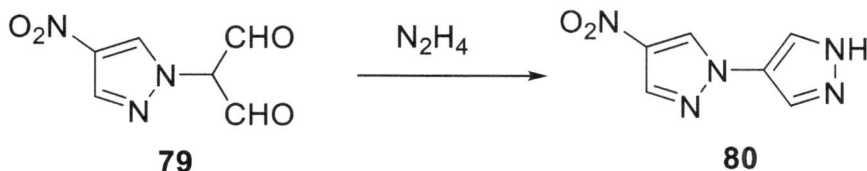

Scheme (23). Synthesis of 1,4`-bipyrazole derivative **80**.

Treatment of 4-bromopyrazole **81** with ethyl bromoacetate or bromoacetonitrile in an anhydrous tetrahydrofuran solution, followed by reaction with dimethyl formamide-diethyl acetal (DMF-DEA) under microwave irradiation condition, resulted in the formation of the 2-(4-bromopyrazol-1-yl)-3-dimethylaminoacrylic acid derivatives **82** (X = CN, CO$_2$Et). Further, microwave irradiation of **82** (X = CN) with hydrazine or phenylhydrazine in ethanol afforded the corresponding 1,4'-bipyrazoles **83** in good yields. For compound **83** (R = H): its ^{1}H NMR in DMSO-d_6 was as following: δ 4.5–6.5 (br s, 2H, NH$_2$), 7.79 (s, 1H, H$_3$), 8.0 (s, 1H, H$_{5'}$), 8.34 (s, 1H, H$_5$), and its ^{13}C NMR displayed six carbon peaks at δ 93.11 (C-Br), 111.44 (C$_{4'}$), 125.27 (C$_{5'}$), 129.96 (C$_5$), 139.98 (C$_3$) and 141.02 (C$_{3'}$) ppm. Similarly, heating of **82** (X = CO$_2$Et) with hydrazine afforded the 1,4'-bipyrazole derivative **84** in excellent yield (Scheme **24**) [53].

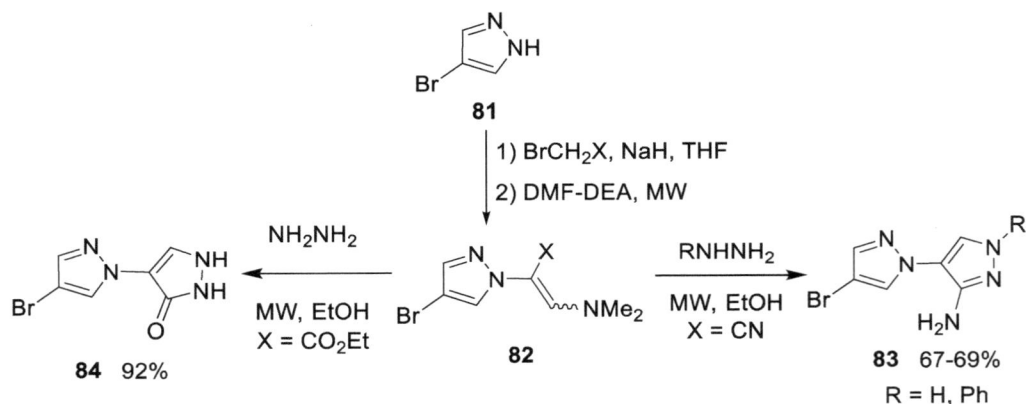

Scheme (24). Synthesis of 1,4`-bipyrazole derivatives **83-84**.

The 1,4`-bipyrazole derivative **89** was synthesized from 3,5-dimethylpyrazole **85** according to the procedure depicted in Scheme (**25**). Thus, treatment of 3,5-dimethylpyrazole **85** with bromoacetonitrile in the presence of *t*BuONa in acetonitrile gave 2-(3,5-dimethyl-1H-pyrazol-1-yl)acetonitrile **86**. Treatment of **86** with ethyl imidazole-1-carboxylate in the presence of NaH in THF afforded compound **87** in high yield. The reaction of **87** with hydrazine hydrate in boiling ethanol afforded compound **88** in excellent yield, which in turn reacted with isopropyl iodide in the presence of K$_2$CO$_3$ to yield the polysubstituted 1,4`-bipyrazole-amine derivative **89** in high yield [16].

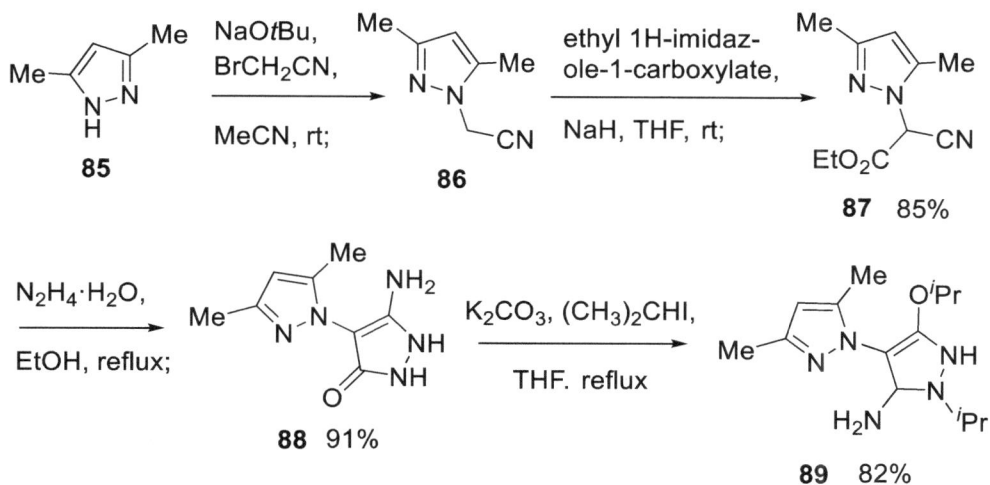

Scheme (25). Synthesis of the 1,4`-bipyrazole derivative **89**.

Treatment of α-(4-chloro-1-pyrazolyl)-4-chloroacetophenone **90** with *bis*(dimethylamino)-methane **91** in dichloromethane at 20-25 °C for 90 min afforded the intermediate **92** which upon reaction with hydrazine hydrate gave the 1,4'-bipyrazole derivative **93** in 58% yield (Scheme **26**) [54].

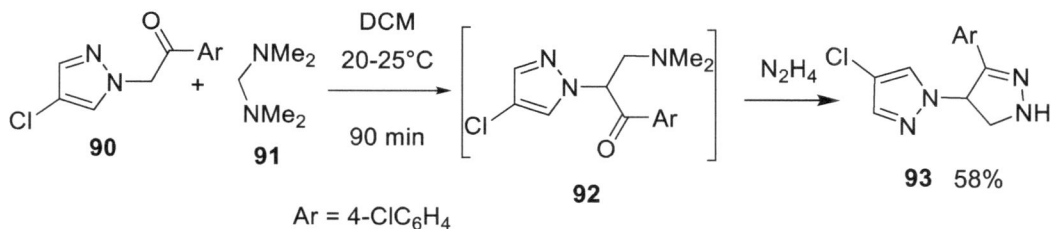

Ar = 4-ClC$_6$H$_4$

Scheme (26). Synthesis of the 1,4`-bipyrazole derivative **93**.

The 2-(1-pyrazolyl)acetophenone derivative **95** was synthesized from alkylation of 3,5-dimethyl-1*H*-pyrazole **85** with phenacyl bromide **94** in refluxing acetone in the presence of potassium carbonate. Condensation of compound **95** with *N,N*-dimethylformamide-dimethylacetal (DMF-DMA) (1.2 equivalent) under reflux gave 3-dimethylamino-2-(3,5-dimethyl-1*H*-1-pyrazolyl)-1-phenyl-2-propen-1-one **96,** that was converted into 1,4'-bipyrazoles **97** by its reaction with hydrazine derivatives (Scheme **27**) [55].

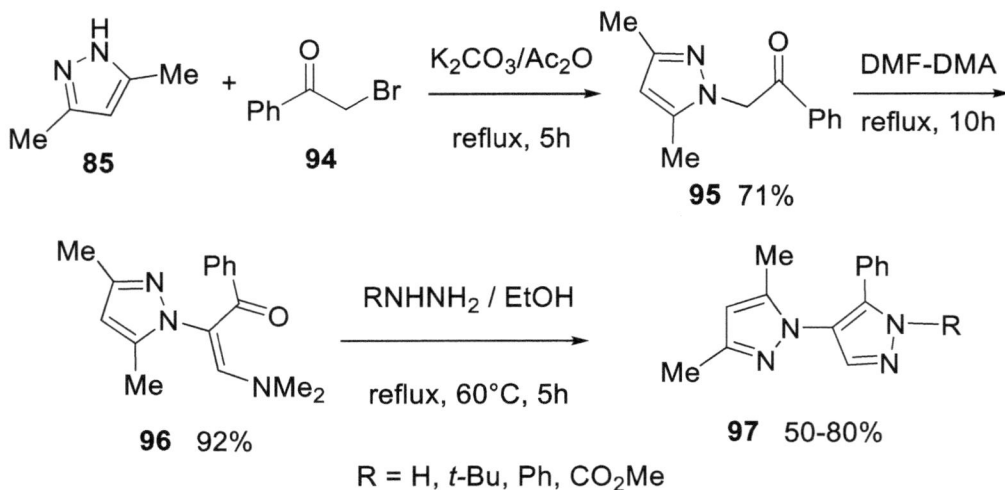

Scheme (27). Synthesis of 1,4`-bipyrazole derivatives **97**.

Reaction of 1*H*-pyrazole-4-carbonitrile **98** with 4-chlorophenacyl bromide **94b** in acetonitrile in the presence of potassium carbonate followed by condensation with formaldehyde resulted in the formation of the pyrazole derivative **99**. Treatment of the latter compound with hydrazine, at refluxing dichloromethane, afforded the corresponding 1,4'-bipyrazole **100**. Reaction of 1,4'-bipyrazole **100** 4-phenylphenylisocyanate afforded the 1,4'-bipyrazole derivative **101** in 69% yield (Scheme **28**) [56, 57].

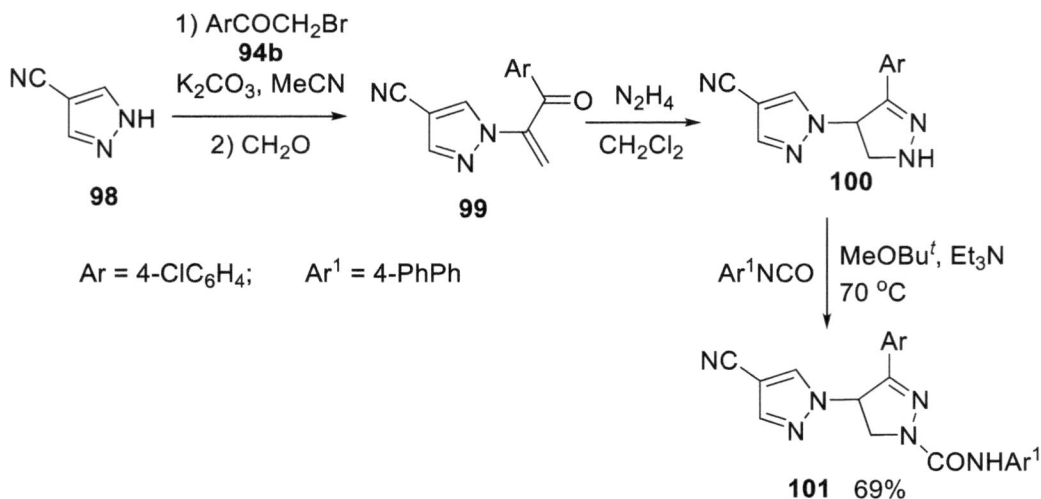

Scheme (28). Synthesis of the 1,4`-bipyrazole derivative **101**.

Treatment of 2-bromo-1,3-diphenylpropanedione **102** with 3-alkylpyrazoles **103** gave the corresponding 2-(pyrazol-1-yl)propanedione derivatives **104** in good yields. When the latter compounds **104** were treated with arylhydrazines in acetic acid/methanol solution, they furnished the 1,4`-bipyrazole derivatives **105** in good yields. Lithiation of **105** followed by treatment with di-alkylchlorophosphine afforded the Bippyphos derivatives **106** in good yields (Scheme **29**) [58 - 60].

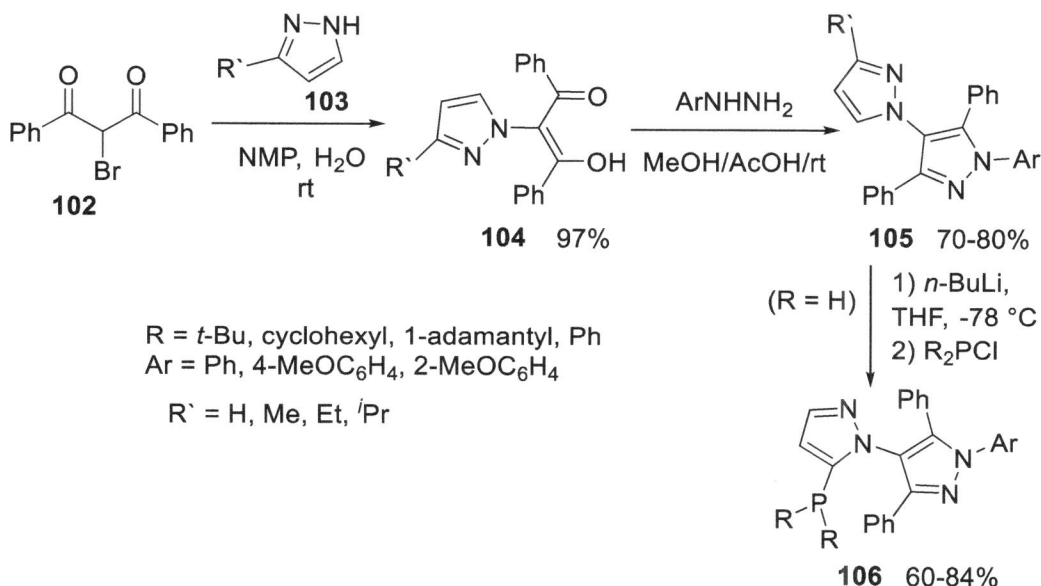

Scheme (29). Synthesis of 1,4`-bipyrazole derivative **106**.

C-*N* Cross coupling reaction was performed between 1*H*-pyrazole **107** with 4-iodo-1-methylpyrazole **49** using cesium carbonate as a base, in the presence of Cu_2O as co-catalyst and salicylaldoxime **108** as a ligand, in acetonitrile to furnish the 1,4'-bipyrazole derivative **109** in 96% yield (Scheme **30**) [61].

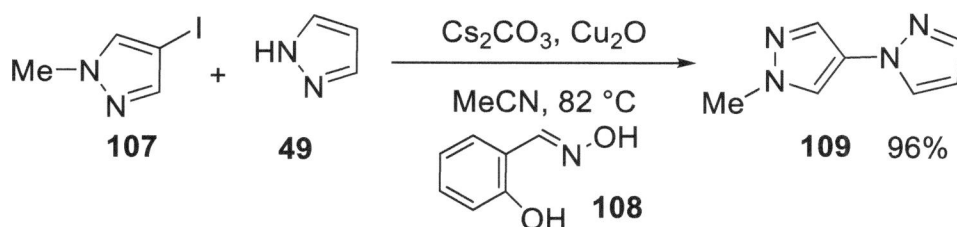

Scheme (30). Synthesis of the 1,4`-bipyrazole derivative **109**.

Reaction of 3,4,5-trinitro-1*H*-pyrazole **110** with 1*H*-pyrazoles **111** in water in the presence of 2 equiv. sodium hydroxide at 80–90 °C, followed by acidification, gave the corresponding 1,4′-bipyrazoles **112** in good yields. Thus, the 1*H*-pyrazoles **111** selectively replaced the 4-positioned nitro group of 3,4,5-trinitro-1*H*-pyrazole **110** (Scheme **31**) [62].

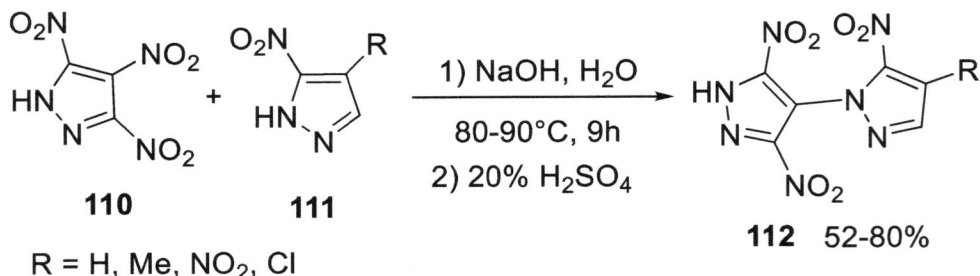

R = H, Me, NO$_2$, Cl

Scheme (31). Synthesis of 1,4`-bipyrazole derivative **101**.

CONCLUSION

In this chapter, we disclosed the various synthetic procedures for the preparation of three classes for bipyrazole systems, namely; 1,1`-bipyrazole, 1,3`-bipyrazole and 1,4`-bipyrazole systems. The most common reactions here were condensation reactions of hydrazines with tetracarbonyl or dihydroxydicarbonyl or pyrazole-based difunctional compounds. In addition, the 1,3-dipolar cycloaddition of pyrazole-based hydrazonoyl halides with activated methylene compounds led to the formation of 1,3`-bipyrazole and 1,4`-bipyrazole derivatives. 1,3`-Bipyrazole and 1,4`-bipyrazole derivatives had potent inhibitory activities against various diseases. Therefore, the reported *N,N-* and *N,C*-bipyrazoles are promising candidates for future research plans in the field of medicinal and material science fields.

REFERENCES

[1]   Ram, V.J.; Sethi, A.; Nath, M.; Pratap, R. The Chemistry of Heterocycles: Nomenclature and Chemistry of Three to Five Membered Heterocycles. Elsevier Ltd., **2019**.

[2]   Raffa, D.; Maggio, B.; Raimondi, M.V.; Cascioferro, S.; Plescia, F.; Cancemi, G.; Daidone, G. Recent advanced in bioactive systems containing pyrazole fused with a five membered heterocycle. *Eur. J. Med. Chem.,* **2015**, *97*, 732-746.
[http://dx.doi.org/10.1016/j.ejmech.2014.12.023] [PMID: 25549911]

[3]   Abdel-Aziz, H.A.; Mekawey, A.A.; Dawood, K.M. Convenient synthesis and antimicrobial evaluation of some novel 2-substituted-3-methylbenzofuran derivatives. *Eur. J. Med. Chem.,* **2009**, *44*(9), 3637-3644.
[http://dx.doi.org/10.1016/j.ejmech.2009.02.020] [PMID: 19321238]

[4]   Dawood, K.M.; Eldebss, T.M.; El-Zahabi, H.S.; Yousef, M.H.; Metz, P. Synthesis of some new pyrazole-based 1,3-thiazoles and 1,3,4-thiadiazoles as anticancer agents. *Eur. J. Med. Chem.,* **2013**, *70*, 740-749.

[http://dx.doi.org/10.1016/j.ejmech.2013.10.042] [PMID: 24231309]

[5]   Dawood, K.M.; Eldebss, T.M.; El-Zahabi, H.S.; Yousef, M.H. Synthesis and antiviral activity of some new bis-1,3-thiazole derivatives. *Eur. J. Med. Chem.,* **2015**, *102*, 266-276.
[http://dx.doi.org/10.1016/j.ejmech.2015.08.005] [PMID: 26291036]

[6]   Dilek Altıntop, M.; Ozdemir, A.; Ilgın, S.; Atli, O. Synthesis and biological evaluation of new pyrazole-based thiazolyl hydrazone derivatives as potential anticancer agents. *Lett. Drug Des. Discov.,* **2014**, *11*(7), 833-839.
[http://dx.doi.org/10.2174/1570180811666140226235350]

[7]   Salameh, B.A.; Abu-Safieh, K.A. AL-Aqrabawi, I.S.; Alsoubani, F.; Tahtamouni, L.H. Synthesis and Cytotoxic Activity of Some New Bipyrazole Derivatives. *Heterocycles,* **2020**, *100*, 283-292.
[http://dx.doi.org/10.3987/COM-20-14222]

[8]   Aggarwal, R.; Sumran, G.; Garg, N.; Aggarwal, A. A regioselective synthesis of some new pyrazol-1--ylpyrazolo[1,5-a]pyrimidines in aqueous medium and their evaluation as antimicrobial agents. *Eur. J. Med. Chem.,* **2011**, *46*(7), 3038-3046.
[http://dx.doi.org/10.1016/j.ejmech.2011.04.041] [PMID: 21558044]

[9]   Veloso, M.P.; Romeiro, N.C.; Silva, G.M.; Alves, H.M.; Doriguetto, A.C.; Ellena, J.; Miranda, A.L.; Barreiro, E.J.; Fraga, C.A. Synthesis and characterization of the atropisomeric relationships of a substituted N-phenyl-bipyrazole derivative with anti-inflammatory properties. *Chirality,* **2012**, *24*(6), 463-470.
[http://dx.doi.org/10.1002/chir.22016] [PMID: 22544569]

[10]   Anderson, K.W.; Fotouhi, N.; Gillespie, P.; Goodnow, R.A., Jr; Guertin, K.R.; Haynes, N.E.; Myers, M.P.; Pietranico-Cole, S.L.; Qi, L.; Rossman, P.L.; Scott, N.R. Preparation of Pyrazole---Carboxamide Derivatives as 11-β-Hydroxysteroid Dehydrogenase Form I (11-β-HSD1) Inhibitors. *PCT Int. Appl,* **2007**. WO 2007107470 A2

[11]   Shigfuji, T. Herbicide composition containing pyrazolylpyrazole compound and other herbicidally-active compound. *Jpn. Kokai Tokkyo Koho,* **2016**. JP 2016141627 A.

[12]   Matsubara, K.; Niino, M. Preparation of pyrazolyl-pyrazole derivatives as herbicides. *PCT Int. Appl,* **2016**, WO 2016117675 A1.

[13]   Matsubara, T.; Niino, M. Substituted pyrazolylpyrazole derivative and its use as herbicide. *PCT Int. Appl.,* **2015**, WO 2015022924 A1.

[14]   Bigot, A.; El Qacemi, M. Preparation of aminocarbonylaryl-substituted (trifluoromethyl)-pyrazolyl sulfonates as agricultural pesticides. *PCT Int. Appl,* **2020**, WO 2020127345 A1.

[15]   Harschneck, T.; Arlt, A.; Velten, R.; Maue, M.; Ilg, K.; Görgens, U.; Tuberg, A. Pyrazoles for controlling arthropods and their preparation. *PCT Int. Appl,* **2018**, WO 2018177993 A1.

[16]   Indrasena Reddy, K.; Aruna, C.; Manisha, M.; Srihari, K.; Sudhakar Babu, K.; Vijayakumar, V.; Sarveswari, S.; Priya, R.; Amrita, A.; Siva, R. Synthesis, DNA binding and *in-vitro* cytotoxicity studies on novel bis-pyrazoles. *J. Photochem. Photobiol. B,* **2017**, *168*, 89-97.
[http://dx.doi.org/10.1016/j.jphotobiol.2017.02.003] [PMID: 28189845]

[17]   Baker-Glenn, C.; Burdick, D.J.; Chambers, M.; Chan, B.K.; Estrada, A; Sweeney, Z.K. (Pyrazolylamino)pyrimidine derivatives as LRRK2 modulators and their preparation and use in the treatment of Parkinson's disease. *PCT Int. Appl,* **2013**, WO 2013164321 A1.

[18]   Gowrisankar, S.; Sergeev, A.G.; Anbarasan, P.; Spannenberg, A.; Neumann, H.; Beller, M. A general and efficient catalyst for palladium-catalyzed C-O coupling reactions of aryl halides with primary alcohols. *J. Am. Chem. Soc.,* **2010**, *132*(33), 11592-11598.
[http://dx.doi.org/10.1021/ja103248d] [PMID: 20672810]

[19]   Kotecki, B.J.; Fernando, D.P.; Haight, A.R.; Lukin, K.A. A general method for the synthesis of unsymmetrically substituted ureas *via* palladium-catalyzed amidation. *Org. Lett.,* **2009**, *11*(4), 947-950.

[http://dx.doi.org/10.1021/ol802931m] [PMID: 19178160]

[20]   Porzelle, A.; Woodrow, M.D.; Tomkinson, N.C. Palladium-catalyzed coupling of hydroxylamines with aryl bromides, chlorides, and iodides. *Org. Lett.,* **2009**, *11*(1), 233-236.
[http://dx.doi.org/10.1021/ol8025022] [PMID: 19035839]

[21]   Yu, S.; Haight, A.; Kotecki, B.; Wang, L.; Lukin, K.; Hill, D.R. Synthesis of a TRPV1 receptor antagonist. *J. Org. Chem.,* **2009**, *74*(24), 9539-9542.
[http://dx.doi.org/10.1021/jo901943s] [PMID: 19928811]

[22]   Beaudoin, D.; Wuest, J.D. Synthesis of N-arylhydroxylamines by Pd-catalyzed coupling. *Tetrahedron Lett.,* **2011**, *52*(17), 2221-2223.
[http://dx.doi.org/10.1016/j.tetlet.2010.12.034]

[23]   Schulz, M.; Mögel, L.; Riediger, W.; Radeglia, R. Radikalreaktionen von N-Heterocyclen. I. Synthese und Struktur eines *N,N*-verknüpften Bipyrazols. *J. Prakt. Chem.,* **1982**, *324*(2), 309-321.
[http://dx.doi.org/10.1002/prac.19823240215]

[24]   De Mendoza, J.; Gonzalez-Muni, M.R.; Martin, M.R.; Elguero, J.N. N'-Linked biazoles. V: Synthesis of pyrazolyl dimers by the reaction of 3-methoxycarbonyl-2-pyrazoline with lead tetraacetate. *Heterocycles,* **1985**, *23*, 2619-2628.
[http://dx.doi.org/10.3987/R-1985-10-2619]

[25]   Tarrago, G.; Ramdani, A.; Elguero, J.; Espada, M. Orientation de la réaction d'alkylation des pyrazoles dans des conditions neutres et en catalyse par transfert de phase. *J. Heterocycl. Chem.,* **1980**, *17*(1), 137-142.
[http://dx.doi.org/10.1002/jhet.5570170128]

[26]   Pfeiffer, W.D.; Bulka, E. Process for preparing 1-pyrazol-3-yl-2-pyrazolin-5-one, Pat. DD 293347A5 (Ger.)

[27]   Ramdani, A.; Tarrago, G. Polypyrazolic macrocycles-I: A study of the polycondensation of 3-chloromethyl-3'(5'),5-dimethyl-5'(3)-pyrazolyl-1-pyrazole. *Tetrahedron,* **1981**, *37*, 987-990.
[http://dx.doi.org/10.1016/S0040-4020(01)97674-4]

[28]   Attayibat, A.; Radi, S.; Ramdani, A.; Lekchiri, Y.; Hacht, B.; Bacquet, M.; Willai, S.; Morcellet, M. Synthesis and cations binding properties of a new C,N-bipyrazolic ligand. *Bull. Korean Chem. Soc.,* **2006**, *27*(10), 1648-1650.
[http://dx.doi.org/10.5012/bkcs.2006.27.10.1648]

[29]   Radi, S.; Attayibat, A.; El-Massaoudi, M.; Bacquet, M.; Jodeh, S.; Warad, I.; Al-Showiman, S.S.; Mabkhot, Y.N. *C,N*-bipyrazole receptor grafted onto a porous silica surface as a novel adsorbent based polymer hybrid. *Talanta,* **2015**, *143*, 1-6.
[http://dx.doi.org/10.1016/j.talanta.2015.04.060] [PMID: 26078121]

[30]   Harit, T.; Isaad, J.; Malek, F. Novel efficient functionalized tetrapyrazolic macrocycle for the selective extraction of lithium cations. *Tetrahedron,* **2016**, *72*(18), 2227-2232.
[http://dx.doi.org/10.1016/j.tet.2016.03.006]

[31]   Khan, M.A.; Freitas, A.C.C. Hetarylpyrazoles. IV(1). Synthesis and reactions of 1,5'-bipyrazoles. *J. Heterocycl. Chem.,* **1983**, *20*(2), 277-279.
[http://dx.doi.org/10.1002/jhet.5570200204]

[32]   Shablykin, O.V.; Brovarets, V.S.; Drach, B.S. New transformations of 5-Hydrazino-2-phenyl-1,3-oxazole-4-carbonitrile. *Russ. J. Gen. Chem.,* **2007**, *77*(5), 936-939.
[http://dx.doi.org/10.1134/S1070363207050210]

[33]   Shablykin, O.V.; Brovarets, V.S.; Rusanov, E.B.; Drach, B.S. Three ways of reactions of 5-(3,--dimethyl-1H-pyrazol-1-yl)-2-phenyl-1,3-oxazole-4-carbonitrile and its analogs with nitrogen-containing bases. *Russ. J. Gen. Chem.,* **2008**, *78*(4), 655-661.
[http://dx.doi.org/10.1134/S1070363208040233]

[34]   Ahmed, N.S.; Saleh, T.S.; El-Mossalamy, E.H. An efficiently sonochemical synthesis of novel

pyrazoles, bipyrazoles and pyrazol-3-ylPyrazolo [3,4-d] pyrimidines incorporating 1H-benzoimidazole. *Curr. Org. Chem.,* **2013,** *17,* 194-202.
[http://dx.doi.org/10.2174/1385272811317020016]

[35]   Hartfiel, U.; Dorfmeister, G.; Franke, H.; Geisler, J.; Johann, G.; Rees, R. Substituted (pyrazolyl)pyrazoles, processes for their preparation and their use as broad-leaf weed herbicides, Eur. *Pat. Appl.,* **1993,** EP 542388 A1 19930519.

[36]   Ren, X.L.; Li, H.B.; Wu, C.; Yang, H.Z. Synthesis of a small library containing substituted pyrazoles. *ARKIVOC,* **2005,** *2005*(xv), 59-67.
[http://dx.doi.org/10.3998/ark.5550190.0006.f09]

[37]   de Mendoza, J.; Prados, P.; Elguero, J. *N,N'*-Linked biazoles. VI: On the structure of compounds derived from the oxidation of 7-methyl-4,5,6,7-tetrahydroindazole. *Heterocycles,* **1985,** *23*(10), 2629-2634.
[http://dx.doi.org/10.3987/R-1985-10-2629]

[38]   Lyalin, B.V.; Petrosyan, V.A.; Ugrak, B.I. Electrosynthesis of 4-chloro derivatives of pyrazole and alkylpyrazoles. *Russ. J. Electrochem.,* **2008,** *44*(12), 1320-1326.
[http://dx.doi.org/10.1134/S1023193508120021]

[39]   Guillou, S.; Janin, Y.L. 5-Iodo-3-ethoxypyrazoles: an entry point to new chemical entities. *Chemistry,* **2010,** *16*(15), 4669-4677.
[http://dx.doi.org/10.1002/chem.200903442] [PMID: 20333718]

[40]   Guillou, S.; Bonhomme, F.J.; Ermolenko, M.S.; Janin, Y.L. Simple preparations of 4 and 5-iodinated pyrazoles as useful building blocks. *Tetrahedron,* **2011,** *67*(44), 8451-8457.
[http://dx.doi.org/10.1016/j.tet.2011.09.029]

[41]   Khan, M.A.; Freitas, A.C. Hetarylpyrazoles III. Synthesis of some 5-azolylpyrazoles. *Monatsh. Chem.,* **1981,** *112*(5), 675-678.
[http://dx.doi.org/10.1007/BF00899681]

[42]   Barreiro, E. J. L.; Fraga, C. A. M.; Palhares de Miranda, A. L.; Rodrigues, C. R.; Veloso, M. P. Preparation of functionalized bipyrazoles as nonsteroidal antiinflammatory agents, Patent Information: Braz. *Pedido PI,* **2001,** BR 9902960, A.

[43]   Berbee, R.P.M.; Habraken, C.L. Pyrazoles. XVIII. Synthesis of a novel tripyrazolyl by two consecutive cine substitution reactions. *J. Heterocycl. Chem.,* **1981,** *18*(3), 559-560.
[http://dx.doi.org/10.1002/jhet.5570180324]

[44]   Cohen-Fernandes, P.; Erkelens, C.; Van Eendenburg, C.G.M.; Verhoeven, J.J.; Habraken, C.L. Synthesis of 3 (5)-(1'-pyrazolyl) pyrazoles from 1,4-dinitropyrazole by cine substitution reaction. Structure determination. *J. Org. Chem.,* **1979,** *44*(23), 4156-4160.
[http://dx.doi.org/10.1021/jo01337a030]

[45]   Dalinger, I.L.; Vatsadze, I.A.; Shkineva, T.K.; Popova, G.P.; Shevelev, S.A. Nucleophilic substitution in 1-methyl-3,4,5-trinitro-1H-pyrazole. *Mendeleev Commun.,* **2011,** *21*(3), 149-150.
[http://dx.doi.org/10.1016/j.mencom.2011.04.012]

[46]   Bruix, M.; Castellanos, M.L.; Martin, M.R.; Mendoza, J. Regioselective nucleophilic substitution in activated 1-aminopyrazolium cations: A facile synthesis of 5-substituted 1-methylpyrazoles. *Tetrahedron Lett.,* **1985,** *26*(44), 5485-5488.
[http://dx.doi.org/10.1016/S0040-4039(00)98243-1]

[47]   Anderson, K.W.; Fotouhi, N.; Gillespie, P.; Goodnow, R.A., Jr; Guertin, K.R.; Haynes, N.; Myers, M.P.; Pietranico-Cole, S.L.; Qi, L.; Rossman, P.L.; Scott, N.R.; Thakkar, K.C.; Tilley, J.W.; Zhang, Q. Pyrazoles as 11-beta-hsd-1 2007, WO 2007107470.

[48]   Mirskova, A.M.; Levkovskaya, G.G.; Voronkov, M.G. Synthesis of biologically active hydrazones and substituted pyrazoles from β,β-dichlorovinyl ketones. *Bull. Acad. Sci. USSR, Div. Chem. Sci.,* **1981,** *30*(6), 1349-1353.

[http://dx.doi.org/10.1007/BF00950298]

[49]   Elnagdi, M.H.; Elmoghayer, M.R.H.; Elfaham, H.A.; Sallam, M.M.; Alnima, H.H. Reactions with heterocyclic amidines. VI. Synthesis and chemistry of pyrazol-5-yl, and 1,2,4-triazol-5-ylhydrazonyl chlorides. *J. Heterocycl. Chem.,* **1980**, *17*(2), 209-212.
[http://dx.doi.org/10.1002/jhet.5570170201]

[50]   Hafez, E.A.A.; Abed, N.M.; Elmoghayer, M.R.H.; El-agamey, A.G.A. Utility of hydrazines and hydrazine derivatives in heterocyclic synthesis. *Heterocycles,* **1984**, *22*(8), 1821-1877.
[http://dx.doi.org/10.3987/R-1984-08-1821]

[51]   Elnagdi, M.H.; Zayed, E.M.; Abdou, S. Chemistry of heterocyclic diazo compounds. *Heterocycles,* **1982**, *19*, 559-578.
[http://dx.doi.org/10.3987/R-1982-03-0559]

[52]   Kral, B.; Kanishchev, M.I.; Semenov, V.V.; Arnold, Z.; Shevelev, S.A.; Fainzilberg, A.A. Synthetic utilization of N-diformylmethylazoles: The preparation of 1-heteryl-4-nitropyrazoles. *Collect. Czech. Chem. Commun.,* **1988**, *53*(7), 1529-1533.
[http://dx.doi.org/10.1135/cccc19881529]

[53]   de la Hoz, A.; Diaz, A.; Elguero, J.; Jimenez, A.; Moreno, A.; Ruiz, A.; Sanchez-Migallon, A. Microwave-assisted synthesis of bipyrazolyls and pyrazolyl-substituted pyrimidines. *Tetrahedron,* **2007**, *63*(3), 748-753.
[http://dx.doi.org/10.1016/j.tet.2006.10.080]

[54]   Gallenkamp, B.; Fuchs, R. Process for the preparation of substituted pyrazolines, Eur. *Pat. Appl.,* **1993**, EP 546420 A1.

[55]   Martins, M.A.; Fiss, G.F.; Frizzo, C.P.; Rosa, F.A.; Bonacorso, H.G.; Zanatta, N. Highly regioselective synthesis of novel 1, 4'-bipyrazoles. *J. Braz. Chem. Soc.,* **2010**, *21*(2), 240-247.
[http://dx.doi.org/10.1590/S0103-50532010000200008]

[56]   Fuchs, R.; Neumann, W. U.; Becker, B.; Erdelen, C.; Stendel, W. Preparation of 4-azolyl-3-phenylpyrazole-1-carboxanilides and analogs as pesticides, Eur. *Pat. Appl,* **1991**, EP 438690 A2.

[57]   Maurer, F.; Fuchs, R.; Turberg, A.; Erdelen, C. Substituted pyrazolines for use as pesticides PCT Int. Appl, **2003**.

[58]   Withbroe, G.J.; Singer, R.A.; Sieser, J.E. Streamlined synthesis of the Bippyphos family of ligands and cross-coupling applications. *Org. Process Res. Dev.,* **2008**, *12*(3), 480-489.
[http://dx.doi.org/10.1021/op7002858]

[59]   Gowrisankar, S.; Sergeev, A.G.; Anbarasan, P.; Spannenberg, A.; Neumann, H.; Beller, M. A general and efficient catalyst for palladium-catalyzed C-O coupling reactions of aryl halides with primary alcohols. *J. Am. Chem. Soc.,* **2010**, *132*(33), 11592-11598.
[http://dx.doi.org/10.1021/ja103248d] [PMID: 20672810]

[60]   Liu, Q.; Gao, Z.; Wei, W.; Lin, X.; Wang, H.; Xu, K.; Shenq, F.Z. Preparation method of compound containing bispyrazole ring and intermediate thereof **2020**.

[61]   Cristau, H.J.; Cellier, P.P.; Spindler, J.F.; Taillefer, M. Mild Conditions for Copper-Catalysed N-Arylation of Pyrazoles. *Eur. J. Org. Chem.,* **2004**, *2004*(4), 695-709.
[http://dx.doi.org/10.1002/ejoc.200300709]

[62]   Dalinger, I.L.; Vatsadze, I.A.; Shkineva, T.K.; Popova, G.P.; Shevelev, S.A. Synthesis of 4-(-azolyl)-3,5-dinitropyrazoles. *Mendeleev Commun.,* **2010**, *20*(6), 355-356.
[http://dx.doi.org/10.1016/j.mencom.2010.11.019]

CHAPTER 2

Chemistry of 3,3`-Bipyrazole Derivatives

Abstract: Synthesis of 3,3`-bipyrazole systems was achieved *via* interesting synthetic methodologies such as 1,3-dipolar cycloaddition reactions, cyclocondensation reactions and metal catalysed C-H activation reactions. Construction of the structurally related 3,3`-bipyrazolines or 3-(pyrazol-3-yl)pyrazolines is described.

Keywords: 3,3`-bipyrazoles, 3,3`-bipyrazolines, Cyclocondensation, Cyclo addition, Cross-coupling, Nitrilimines.

1. INTRODUCTION

3,3`-Bipyrazoles, 3,3`-bipyrazolines and 3-(pyrazol-3-yl)pyrazolines are all structurally related C-C directly connected two pyrazole units by sigma bond between 3,3`-positions without any spacer. There are several tautomeric structural formulae that can be drawn for such 3,3`-bipyrazole derivatives, as depicted in Scheme (**1**). The synthetic pathways for the 3,3`-bipyrazole structures are briefly summarized in Scheme (**2**). Such routes are: 1) reactions of tetracarbonyl or dihydroxydicarbonyl building units with hydrazines, 2) reaction of pyrazoles having a difunctional-side arm at position 3 with hydrazines, 3) reaction of 3-pyrazolylhydrazines with difunctional compounds, 4) 1,3-dipolar cycloaddition of pyrazolyl-nitrilimines with olefins or acetylenes, and 5) 1,3-dipolar cycloaddition of bis-nitrilimines with two equivalents of olefins or acetylenes.

The 3,3`-bipyrazole derivatives had several academic and industrial applications. They formed complexes with copper(I/II) that were efficiently used for oxidation of catechol to o-quinine with the atmospheric dioxygen [1]. Their ruthenium(II) complexes showed good catalytic activity and transfer of hydrogen in catalyzed hydrogenation reactions [2, 3], and their palladium(II)-complexes were reported as good precatalysts for Suzuki-Miyaura C-C cross-coupling reactions in aqueous media [4]. They have involved in the synthesis of poly(3,3′-bipyrazole) derivatives with high thermal stability and electrochemical activity [5]. Nitration of 3,3-bipyrazole gave several polynitro-3,3`-bipyrazole derivatives that were found to be metal-free primary explosives with high energetic properties and excellent thermal stability [6 - 8]. .

Kamal M. Dawood and Ashraf A. Abbas

The 3,3'-bipyrazole derivatives also had solvatochromic behaviour [9] The platinum and osmium complexes of 3,3`-bipyrazoles were also useful as emitting materials for organic light-emitting diode (OLED) [10 - 12]. 3,3`-Bipyrazole derivatives were also reported to have high antitumor inhibitory activity [13].

Scheme (1). The possible tautomeric forms of 3.3`-bipyrazoles.

Scheme (2). The possible synthetic routes to 3,3`-bipyrazoles.

2. SYNTHESIS OF 3,3`-BIPYRAZOLE SYSTEMS

2.1. From 1,3-Dipolar Cycloaddition Reactions

When the *bis*-arylnitrilimines **2** (generated *in situ* from the treatment of *bis*-hydrazonyl halides **1** with triethylamine in dry benzene) was treated with the active methylene compounds **3**, they resulted in the formation of the 3,3'-bipyrazole derivatives **4** in high yields. Similarly, the *bis*-arylnitrilimines **2** underwent 1,3-dipolar cycloaddition reactions with the activated olefins **5** to give the 3,3'-bi(2-pyrazolines) **6**. Oxidation of compound **6** (R^2 = Ph, R^3 = COPh, Ar = Ph) with chloranil afforded the corresponding 3,3'-bipyrazole derivative **7** in 71% yield (Scheme **3**) [14].

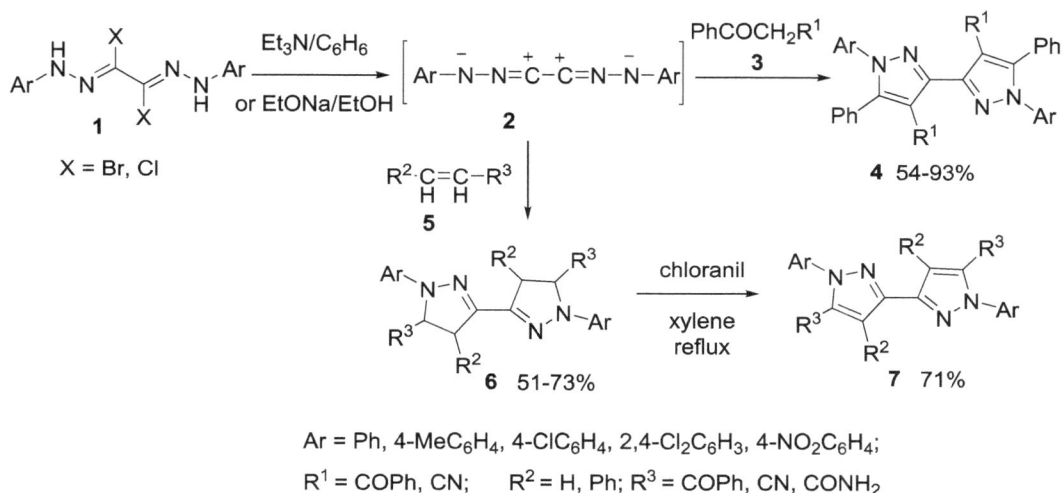

Scheme (3).   Synthesis of 3,3`-bipyrazole **4** and **7**.

Regioselective synthesis of polysubstituted 3,3'-bi-1*H*-pyrazole derivatives **10** was carried out *via* 1,3-dipolar cycloaddition reaction of the *bis*-arylnitrilimines **2** with the cinnamonitriles **8** to yield the cycloadducts 5,5'-dicyano-4,4',5,5'-tetrahydro-3,3'-bi-1*H*-pyrazoles **9** in 40-75% yields. Aromatization of compounds **9** *via* thermal elimination of hydrogen cyanide under the basic reaction conditions afforded the 3,3'-bi-1*H*-pyrazole derivatives **10** in good yields (Scheme **4**) [15].

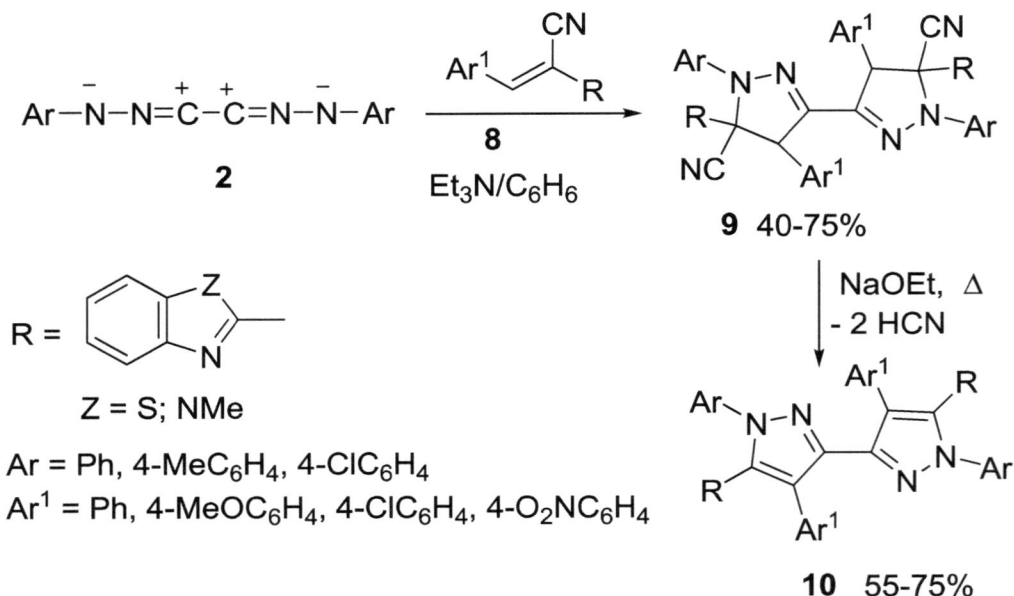

Scheme (4). Synthesis of 3,3`-bipyrazole **10**.

When the *bis*-hydrazonoyl halide **1** was treated with fumaronitrile **11** in benzene under reflux condition in the presence of triethylamine as a base, it furnished 1,1'-diphenyl-3,3'-bipyrazole-4,4'-dicarbonitrile **13** in moderate yield. The cycloaddition reaction proceeded *via* the loss of two molecules of hydrocyanic acid (HCN) from the intermediate **12** (Scheme **5**) [16].

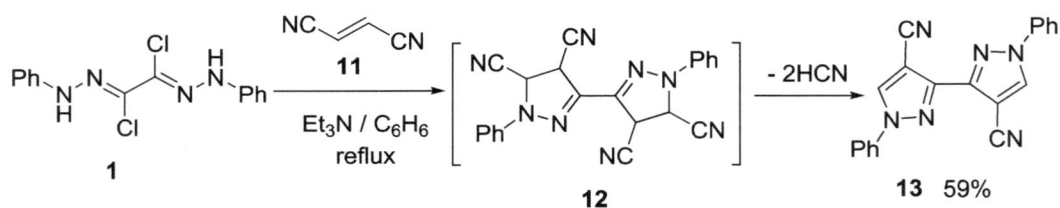

Scheme (5). Synthesis of 3,3`-bipyrazole **13**.

In a similar pathway, the *bis*-nitrilimines **2** reacted with the acetylenecarboxylates **14,** in 1:2 molar ratios, in benzene at reflux temperature using triethylamine as a base. The reaction products 1,1'-aryl-3,3`-bipyrazole-4,4'-dicarboxylates **15** were obtained in good yields *via* regioselective 1,3-dipolar cycloaddition. Furthermore, *bis*-hydrazonoyl chlorides **2** reacted with dimethyl acetylenedicarboxylate (**16**) and gave the tetramethyl 1,1'-diaryl-3,3`-bipyrazole-4,4`,5,5`-tetracarboxylate esters **17** in acceptable yields. Heating either the cycloadducts **15a** and **17a** with a mixture of HCl/AcOH, led to the formation of the same product; 3,3`-bipyrazol--5,5`-dicarboxylic acid **18**. When the tetracarboxylate ester **17** was treated with hydrazine and with aniline derivatives in refluxing dimethylformamide (DMF), it afforded the bipyrazole-fused heterocyclic systems **19** and **20**, respectively, (Scheme 6) [4, 17]. The ^{1}H NMR of compound **19** in DMSO-d_6 displayed multiplet aromatic-H peaks at δ 7.63-7.53 and a singlet peak at 7.92 for NH protons, and its ^{13}C-NMR showed peaks at δ 161.45, 160.09, 143.29, 138.47, 131.49, 129.47, 128.77,123.78, 114.83 ppm.

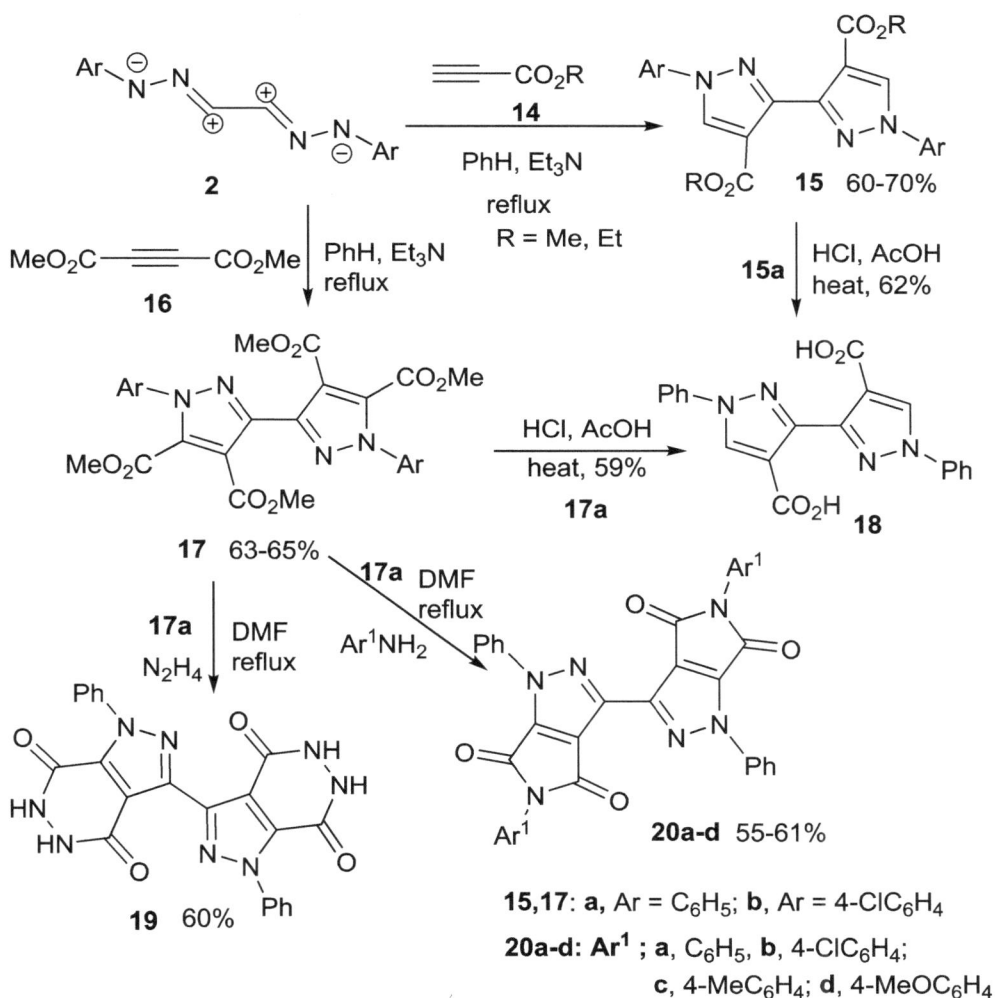

Scheme (6). Synthesis of the 3,3`-bipyrazole derivatives **19-20a-d**.

Next, 1,3-dipolar cycloaddition of the *bis*-hydrazonoyl halides **1** with the benzylidene chroman-4-one and thiochroman-4-one derivatives **21** proceeded regioselectively to give the corresponding *bis*-spiropyrazoline-5,3'-chroman(thiochroman)-4-one derivatives **22** in good yields. X-ray single crystal analysis confirmed the regio- and stereo- selectivity of the cycloaddition pathway. *Bis*-hydrazonoyl halides **1** reacted similarly with 2-benzylidene-3-coumaranone **23** to afford the 3,3'-bipyrazole derivatives **24** in moderate yields (Scheme 7) [18].

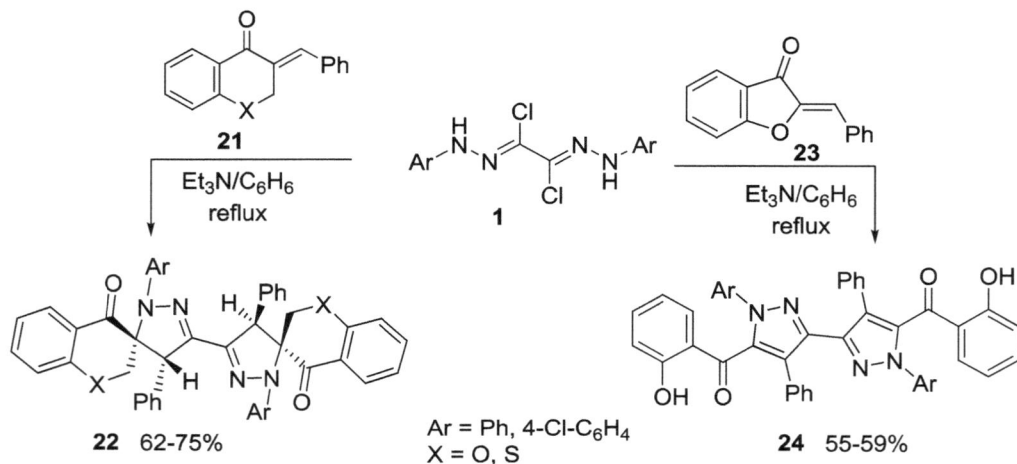

Scheme (7). Synthesis of the 3,3'-bipyrazole derivatives **22,24**.

In the same fashion, the *bis*-nitrilimine **2** underwent a double 1,3-dipolar cycloaddition reaction with two equivalents of the 2-arylidene-1-benzosuberone derivatives **25** to afford the *bis*-[spiro-benzosuberane-2,5'-pyrazoline] derivatives **26** (Scheme 8). The reaction was conducted under both ultrasonic irradiation conditions (at 70 °C for 3h) as well as under conventional thermal heating mode (at reflux for 36 h) in ethanol using triethylamine as a base. Single crystal X-ray analysis confirmed exclusively the regio- and stereo-selectivities of the bipyrazole adducts **26**. The reaction proceeded more efficiently under sonication (72-91% yields) compared with conventional heating (22-51% yields) [19].

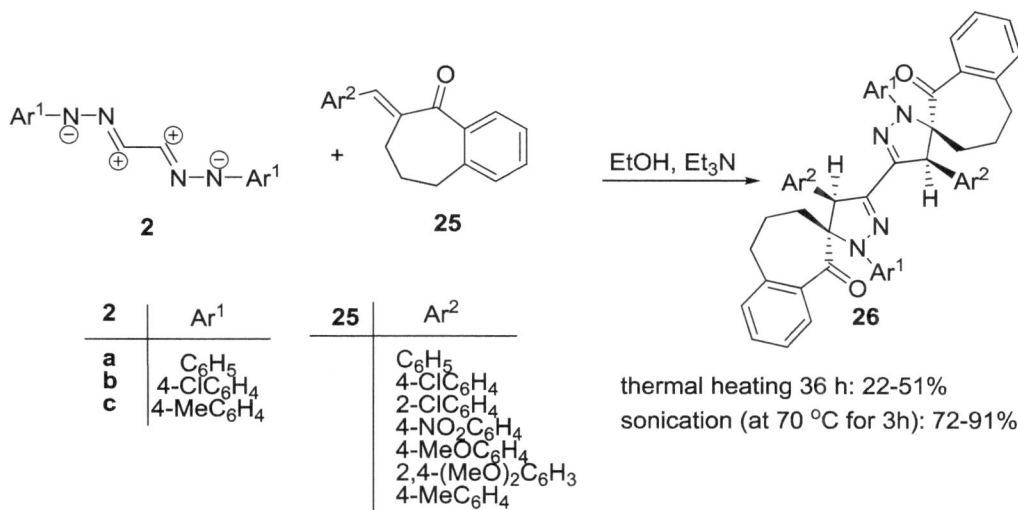

2	Ar^1	25	Ar^2
a	C$_6$H$_5$		C$_6$H$_5$
b	4-ClC$_6$H$_4$		4-ClC$_6$H$_4$
c	4-MeC$_6$H$_4$		2-ClC$_6$H$_4$
			4-NO$_2$C$_6$H$_4$
			4-MeOC$_6$H$_4$
			2,4-(MeO)$_2$C$_6$H$_3$
			4-MeC$_6$H$_4$

thermal heating 36 h: 22-51%
sonication (at 70 °C for 3h): 72-91%

Scheme (8). Synthesis of the 3,3`-bipyrazole derivatives **26**.

The 1,3-dipolar cycloaddition reaction of the 4-arylidene-pyrazolin-5-one derivatives **27** with bis-hydrazonoyl chlorides **1** in 2:1 molar ratios was conducted under conventional thermal heating and ultrasonic irradiation conditions in ethanol and triethylamine as a base. The reaction resulted in the formation of the bis-[3-methyl-1,1`,4`-triaryl-5-oxo-spiro-pyrazoline-3,4`-pyrazoline] derivatives **28** as outlined in Scheme (**9**). The reaction was highly regio- and stereoselective and the ultrasonic irradiation condition was superior to conventional heating one where the reactions were complete within 3 hours under ultrasonic condition compared with 36 hours under conventional heating. In addition, the reaction yields ranged between 72-90% under ultrasonic conditions compared with 19-35% under conventional heating mode [20].

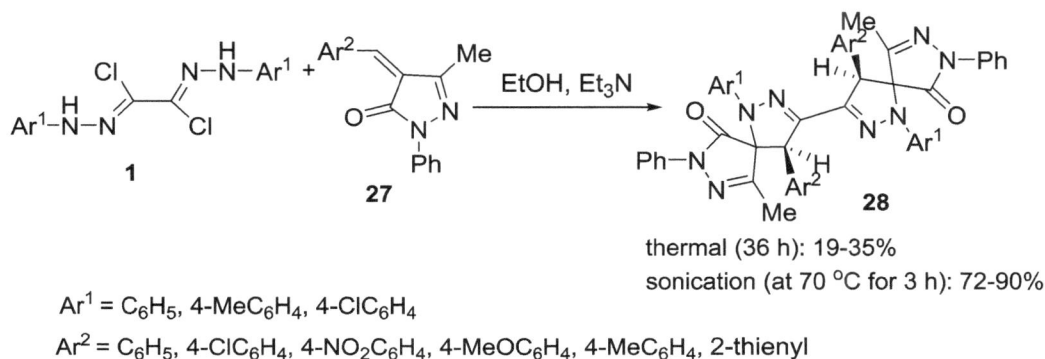

thermal (36 h): 19-35%
sonication (at 70 °C for 3 h): 72-90%

Ar^1 = C$_6$H$_5$, 4-MeC$_6$H$_4$, 4-ClC$_6$H$_4$

Ar^2 = C$_6$H$_5$, 4-ClC$_6$H$_4$, 4-NO$_2$C$_6$H$_4$, 4-MeOC$_6$H$_4$, 4-MeC$_6$H$_4$, 2-thienyl

Scheme (9). Synthesis of the 3,3'-bipyrazole derivatives **28**.

Reaction of 5-pyrazolylformylhydrazones **29** with β-nitrostyrenes **30** under microwave irradiation at 130 °C and solvent free condition led to the formation of a mixture of the 3,3'-bipyrazoles **31** and **32** *via* 1,3-dipolar cycloaddition. The regioselectivity of the obtained product was discussed according to the supposed mode of addition outlined in Scheme (**10**) followed by elimination of HNO_2 to give the aromatic bipyrazoles **31** and **32** [21, 22].

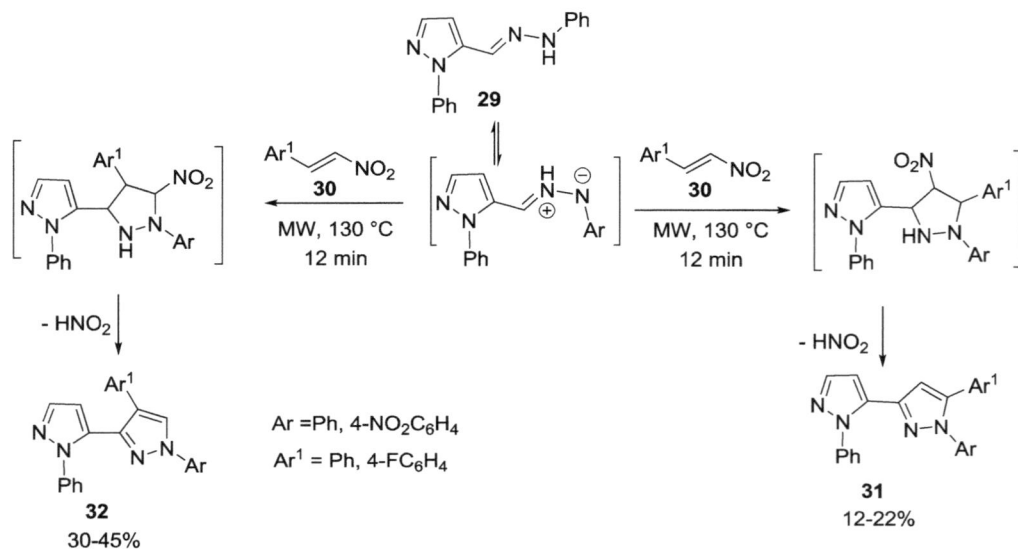

Scheme (10). Synthesis of the 3,3'-bipyrazole derivatives **31** and **32**.

3,3'-Bipyrazole derivatives **35** and **38** were prepared in reasonable yields *via* 1,3-dipolar cycloaddition of 5-pyrazolylformylhydrazone **29** with the electron poor dipolarophiles; dimethyl fumarate **33** and ethyl 3-phenylpropiolate **36**, respectively, under microwave irradiation condition. Formation of the bipyrazole derivatives **35** and **38** were assumed *via* the dipolar intermediates **34** and **37** which was then aromatized *in situ via* air oxidation (Scheme **11**) [21].

The 1,3-dipolar cycloaddition of the activated olefins **40** with 4-pyrazolylhydrazonoyl bromides **39** proceeded smoothly and furnished the corresponding unsaturated 3,3'-bipyrazole derivatives **41** regioselectively. Similar reaction of the hydrazonoyl bromides **39** with acetylenedicarboxylate ester **42** gave the 3,3'-bipyrazole-diesters **43** (Scheme **12**) [23].

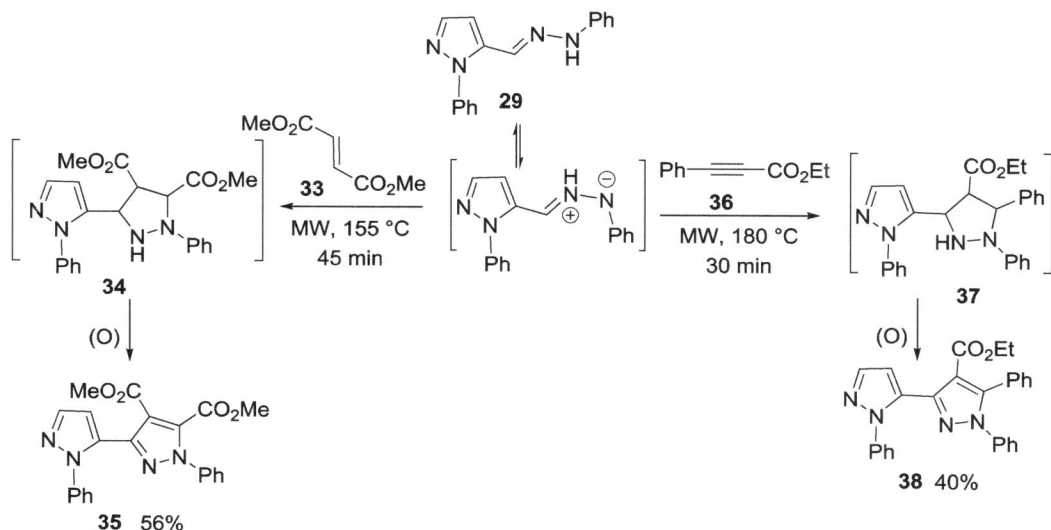

Scheme (11). Synthesis of the 3,3'-bipyrazole derivatives **35**, **38**.

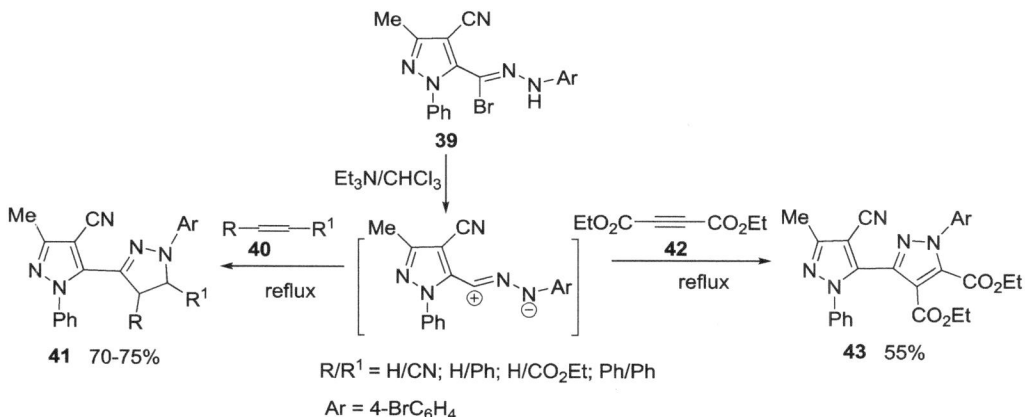

Scheme (12). Synthesis of the 3,3'-bipyrazole derivatives **41**, **43**.

1,3-Dipolar cycloaddition of 3-styrylpyrazoles **46** with the nitrilimines **45**, obtained *in situ* from the hydrazonyl chloride **44**, in benzene at reflux, using triethylamine as a base, afforded the respective 3,3'-bipyrazole derivatives **47** in high yields (Scheme **13**) [24].

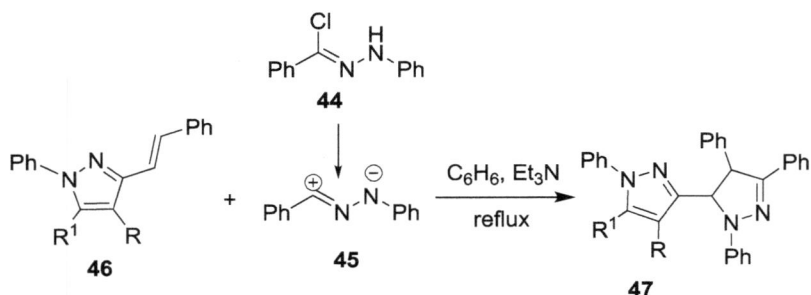

R^1 / R = H/CN, Ph/CN, COMe/Me, CO$_2$Et/Me, CO$_2$Et/Ph, CN/Ph, COPh/Ph

Scheme (13). Synthesis of the 3,3'-bipyrazole derivatives **47**.

Treatment of 3-cyanoacetylpyrazole **49** with hydrazonoyl chlorides **48** in an ethanolic solution of sodium ethoxide at room temperature led to the production of the 3,3'-bipyrazole derivatives **51**. The reaction pathway was assumed to proceed *via* loss of water from the intermediate **50** (Scheme **14**) [25].

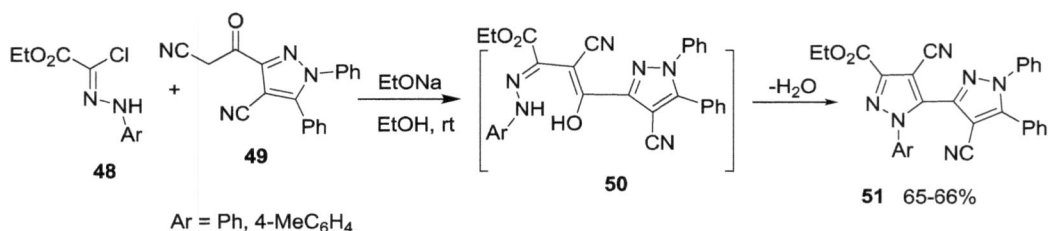

Ar = Ph, 4-MeC$_6$H$_4$

Scheme (14). Synthesis of the 3,3'-bipyrazole derivatives **51**.

1,3-Dipolar cycloaddition reaction of 2,3-*bis*(phenylsulfonyl)-1,3-butadiene **52** with diazomethane in dichloromethane was carried out at ambient temperature under nitrogen atmosphere to give a 1:1 mixture of the pyrazoline **53** and 3,3'-bipyrazoline **54** derivatives. The 3,3'-bipyrazoline **54** underwent a thermal extrusion of nitrogen to produce (*E,E*)-3,4-diphenylsulfonylhexadiene **55** (Scheme **15**) [26].

When 2-phenylsulfinyl-3-phenylsulfonyl-1,3-butadiene **52** was allowed to react with an excess of diazomethane for longer reaction times, the dihydro-3,3'-bipyrazole **57** was obtained solely in good yield. Formation of compound **57** was proceeded *via* 1,3-dipolar cycloaddition reaction of diazomethane with both π-bonds in a sequential manner giving the 2:1-adduct **52** as an intermediate which underwent a subsequent *syn* elimination of PhSOH then a 1,5-sigmatropic hydrogen shift to give **57** (Scheme **16**) [26].

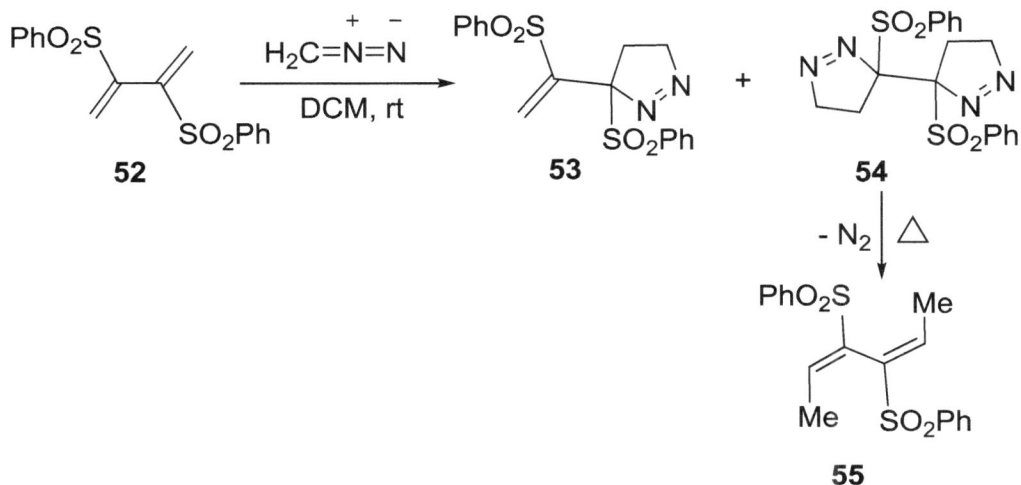

Scheme (15). Synthesis of the 3,3'-bipyrazole derivatives **53,54**.

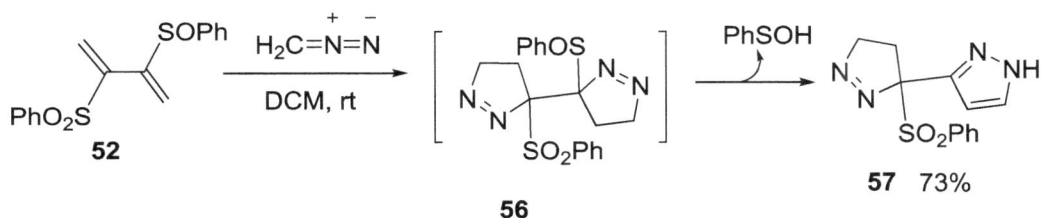

Scheme (16). Synthesis of the 3,3'-bipyrazole derivatives **57**.

1,3-Dipolar cycloaddition reaction of buta-1,3-diyne **58** with 2-diazopropane **59** was performed in two steps in diethyl ether to afford the acetylenic pyrazole **60**. Further cycloaddition process of **60** with 2-diazopropane **59** at 0°C to afford the 5,5,5',5'-tetramethyl-3,3'-bipyrazole **61** in an acceptable yield. The ^{1}H NMR of compound **61** showed two singlet peaks at δ 1.54 and 7.53 ppm for four CH$_3$ and two CH groups. The latter bipyrazole **61** was converted into 2,7-dimethylocta-2-6-dien-4-yne **62,** *via* loss of two N$_2$ molecules, under photolysis condition (Scheme **17**) [27].

The 3,3`-bipyrazole derivatives **65** were synthesized regioselectively in high yields *via* a one-pot three-component protocol by reaction of the aroylideneketene dithioacetals **63** with Bestmann Ohira reagent **64** and hydrazine hydrate using KOH as a base in methanol solvent (Scheme **18**). The sequence of the reaction was suggested to proceed through 5 steps: 1,3-dipolar cycloaddition, protonation, air oxidation, pyrazole formation and demethylation, respectively, according to the proposed mechanism presented in Scheme (**19**) [28].

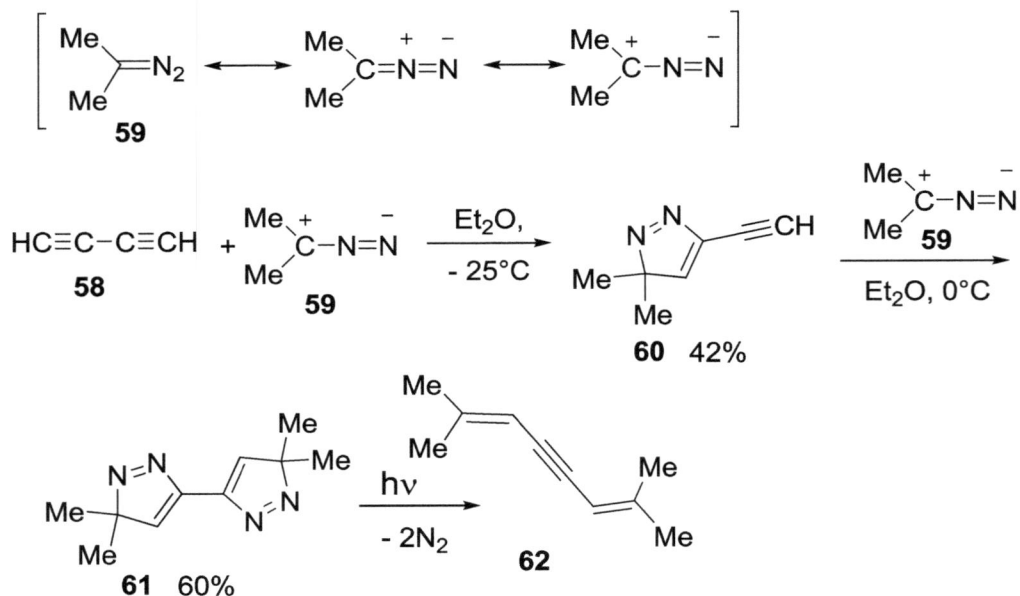

Scheme (17). Synthesis of the 3,3'-bipyrazole derivatives **61**.

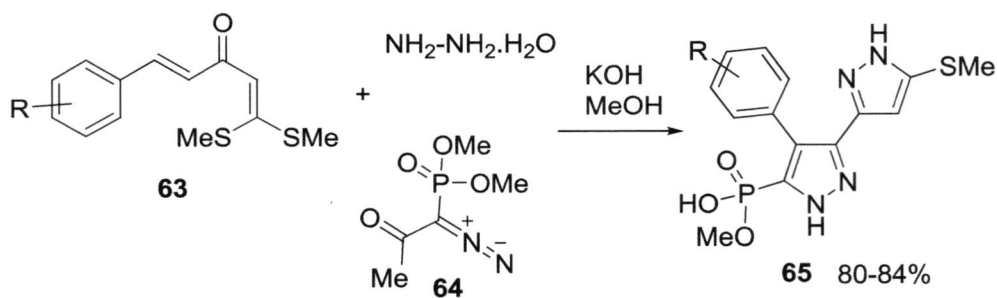

R = 4-Br,4-Cl, 4-F, 4-Me, 2-Me,4-OMe,
4-OEt, 2-OMe, H, 2-F, 2,4-Cl$_2$, 3-Br and 4-F,
1-naphthyl, 5-anthracneyl, 2-thienyl

Scheme (18). Synthesis of the 3,3'-bipyrazoline derivatives **65**.

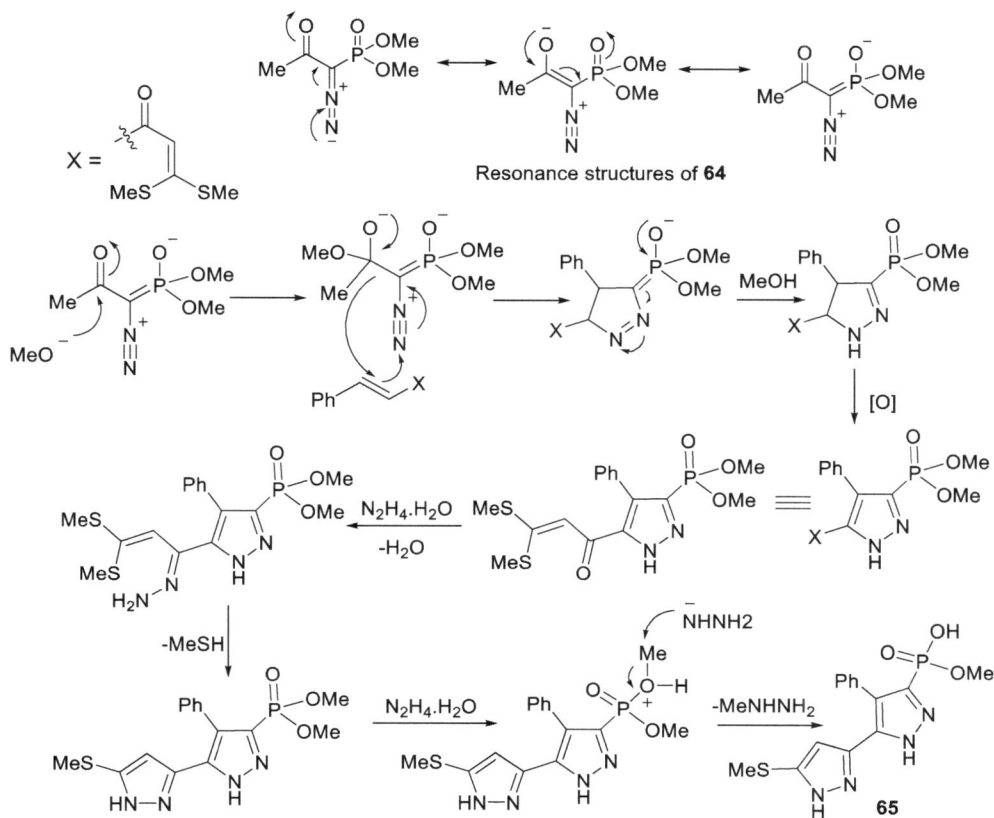

Scheme (19). A proposed mechanism for the formation of **65**.

2.2. *Via* Cyclocondensation Reactions

1,6-Di(2-pyridyl)-1,3,4,6-hexanetetrone **67** was obtained from reaction of 2-acetylpyridine **66** with diethyl oxalate in the presence of sodium ethoxide, then condensation reaction of **67** with hydrazine hydrate afforded 5,5'-di(2-pyridyl)-3,3'-bipyrazole **68** (Scheme **20**) [29].

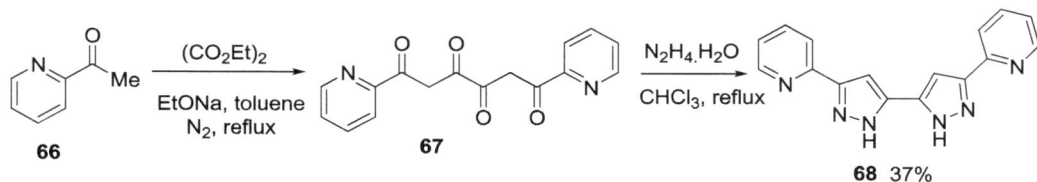

Scheme (20). Synthesis of the 3,3'-bipyrazole derivative **68**.

Treatment of the 1,3,4,6-tetraketones **69** with hydrazine hydrate in ethanol at refluxing condition afforded the corresponding 3,3'-bipyrazole derivatives **70**. *N,N'*-arylation of the latter compounds **70** with activated fluorobenzenes under conventional heating mode, afforded the 1,1´-diaryl-3,3´-bipyrazole derivatives **71** (Scheme **21**) [11, 61, 62]. The bipyrazole derivatives **70** underwent alkylation when treated with benzylbromide and ethyl chloroacetate in refluxing THF to give the corresponding alkylated 3,3´-bipyrazole derivatives **72** in moderate to good yields (Scheme **21**) [2, 30 - 35].

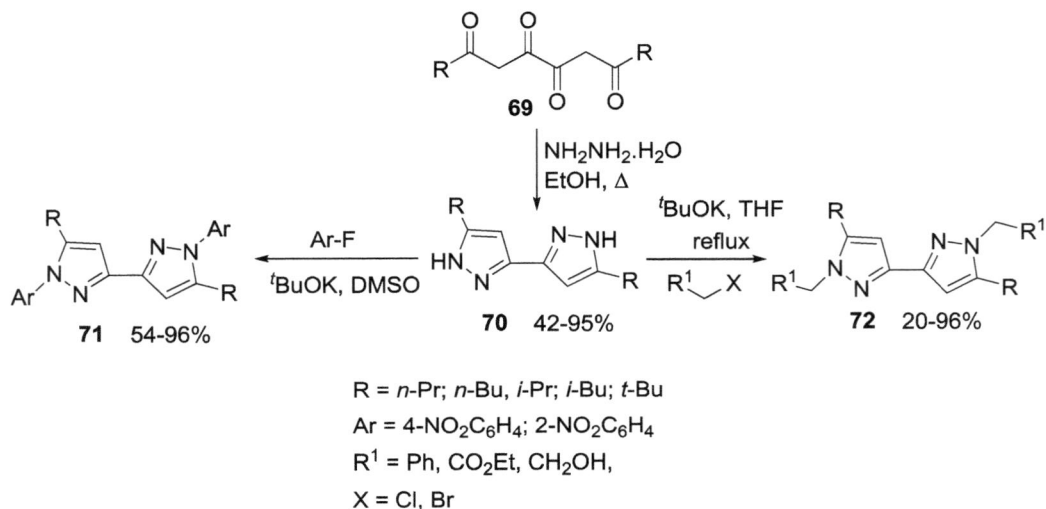

R = *n*-Pr; *n*-Bu, *i*-Pr; *i*-Bu; *t*-Bu

Ar = 4-NO$_2$C$_6$H$_4$; 2-NO$_2$C$_6$H$_4$

R^1 = Ph, CO$_2$Et, CH$_2$OH,

X = Cl, Br

Scheme (21). Synthesis of the 3,3'-bipyrazole derivative **71,72**.

Heating the tetraketone derivatives **73** with hydrazines in refluxing ethanol resulted in the production of the corresponding 3,3'-bipyrazole derivatives **74** in reasonable yields (Scheme **22**) [36 - 38].

Condensation of the tetraketone derivative **69** with hydrazine hydrate in ethanol at reflux gave the 1*H*,1`*H*-3,3`-bipyrazole derivative **75**. Heating the bipyrazole **75** with 4-fluoronitrobenzene in the presence of potassium *tert*-butoxide in DMSO produced the 1,1'-di(4-nitrophenyl)-3,3'-bipyrazole **76**. Reduction of the nitro groups of **76** using Sn/HCl mixture afforded the 1,1`-di(4-aminophenyl)-5-5`-diisopropyl-3,3`-bipyrazole (**77**) in 95% yield (Scheme **23**) [39].

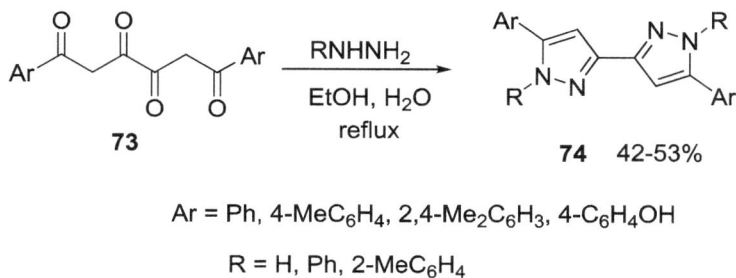

Ar = Ph, 4-MeC$_6$H$_4$, 2,4-Me$_2$C$_6$H$_3$, 4-C$_6$H$_4$OH

R = H, Ph, 2-MeC$_6$H$_4$

Scheme (22). Synthesis of the 3,3'-bipyrazole derivative **74**.

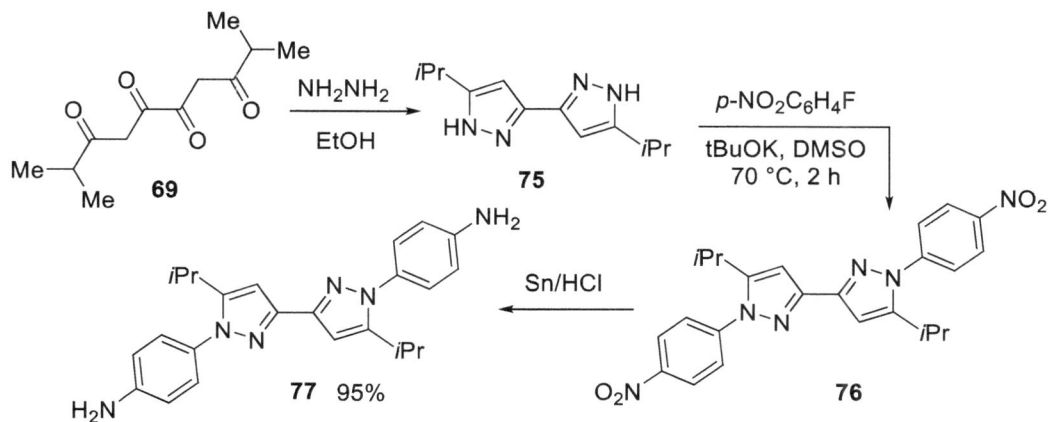

Scheme (23). Synthesis of 5,5'-diisopropyl-3,3'-bipyrazoles **77**.

The 3,3`-bi(camphopyrazole) derivative **80** was synthesized utilizing dihydroxybutadiene derivative **78** with α-naphthylhydrazine hydrochloride **79** in hot ethanol then the reaction mixture was neutralized with a saturated solution of sodium carbonate (Scheme **24**) [40].

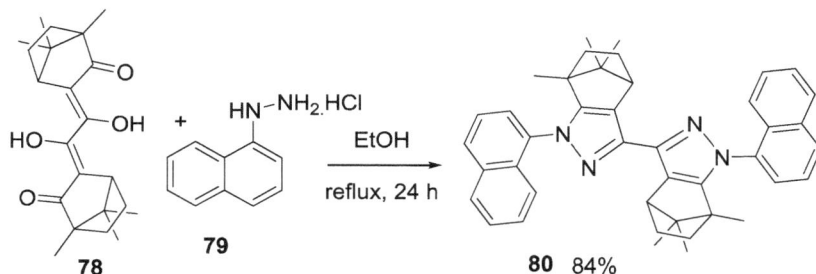

Scheme (24). Synthesis of the 3,3'-bipyrazole derivative **80**.

The reaction of pyrazole **81** with acetone in the presence of sodium metal gave the pyrazole derivative **82**. The reaction of hydrazine hydrate with the pyrazole derivative **82,** in MeCN and ethanol at room temperature, afforded 1,5,5'-trimethyl-1H,1'H-3,3'-bipyrazole (**83**) in 90% yield as shown in Scheme (**25**) [41, 42].

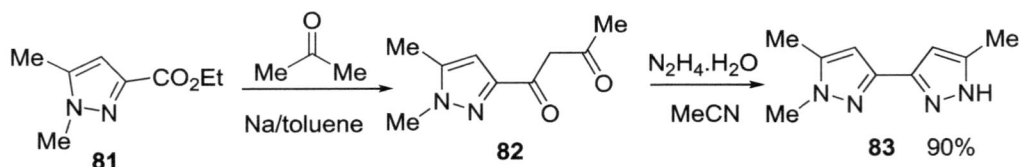

Scheme (25). Synthesis of the 3,3'-bipyrazoline derivatives **83**.

Treatment of the *bis*-(benzoyl)buta-1,3-diyne **84** with methyl hydrazine in dichloromethane afforded the 3,3'-bipyrazole derivative **85** in a good yield (Scheme **26**). However, the reaction of compound **84** with phenylhydrazine produced a mixture of the three regioisomeric *bis*-pyrazoles **86**, **87** and **88**, in a 1:1.5:0.5 ratios, as determined by ¹H NMR. In addition, lithiation of the pyrazolylacetylene derivatives **89** and **90** with butyl lithium followed by quenching of the acetylides with acetic anhydride gave the corresponding ketones **91** and **92**. Heating the acetylenic ketones **91** and **92** with phenylhydrazine hydrochloride and potassium carbonate in methanol afforded the bipyrazoles **93** and **94** in reasonable yields (Scheme **26**) [43].

The 3,3'-bipyrazole derivatives **97** were obtained in good yields by treating (1,5-diaryl-2-hydroxy-3-oxopyrrolidin-2-yl)acetates **95** with hydrazine hydrate in ethanol at reflux. The product **95** was obtained *via* loss of aniline derivative and water. The 3.3'-bipyrazole derivatives **97** were alternatively synthesized by hydrazinolysis of the 5-aryl-2-alkoxycarbonylmethylene-2,3-dihydro-3-furanones **98**. Mechanistically, the formation of compounds **97** was supposed to proceed *via* the reaction of hydrazine with the 6-aryl-6-arylamino-3,4-dioxo-5-hexenoic acids acyclic esters **96**, which were the oxo-form of the starting compounds **95** (Scheme **27**) [44, 45].

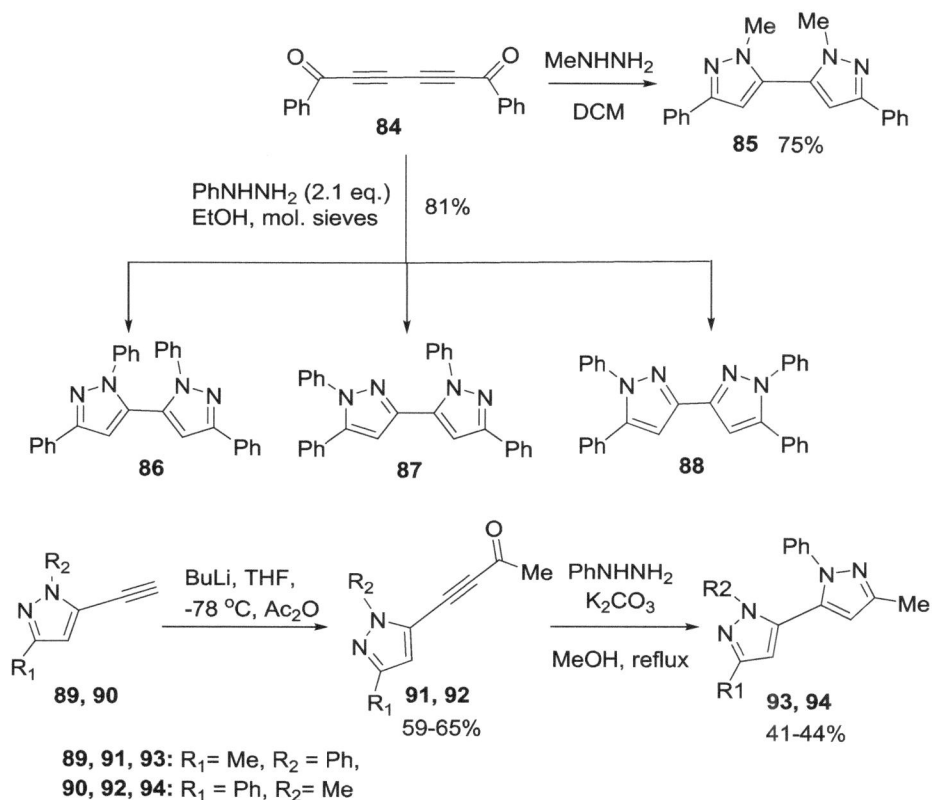

Scheme (26). Synthesis of the 3,3'-bipyrazoline derivatives **86-88**, **93,94**.

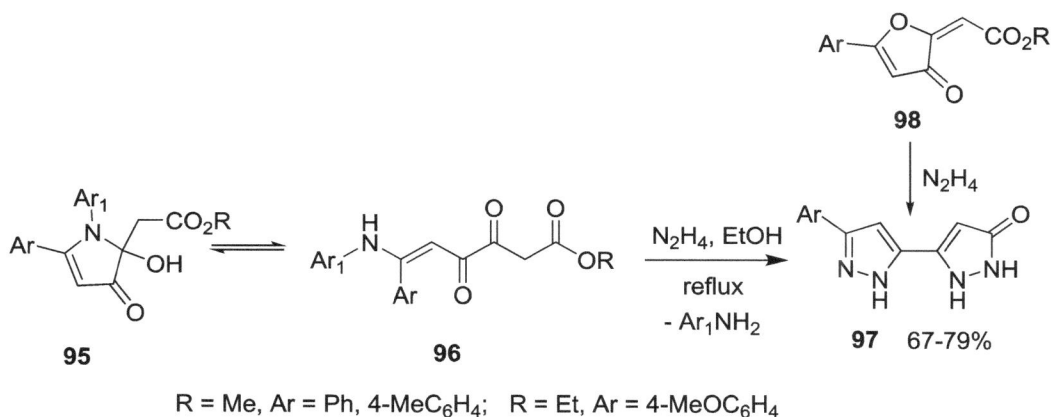

R = Me, Ar = Ph, 4-MeC$_6$H$_4$;   R = Et, Ar = 4-MeOC$_6$H$_4$

Scheme (27). Synthesis of the 3,3'-bipyrazoline derivatives **97**.

Preparation of 5,5'-bis(trifluoromethyl)-3,3'-bipyrazole **102** was conducted employing acetyl pyrazole **99** as outlined in Scheme (**28**). Thus, when acetyl pyrazole **99** was treated with ethyl trifluoroacetate using LiH in THF solvent, it afforded the lithium salt **100** in 89% yield, which upon heating with hydrazine hydrochloride in methanol furnished the pyrazolinyl-pyrazole derivative **101** in 70% yield. Heating the latter derivative **101** in chlorobenzene resulted in the loss of water to afford the aromatic 3,3'-bipyrazole derivative **102** in 92% yield [46, 47].

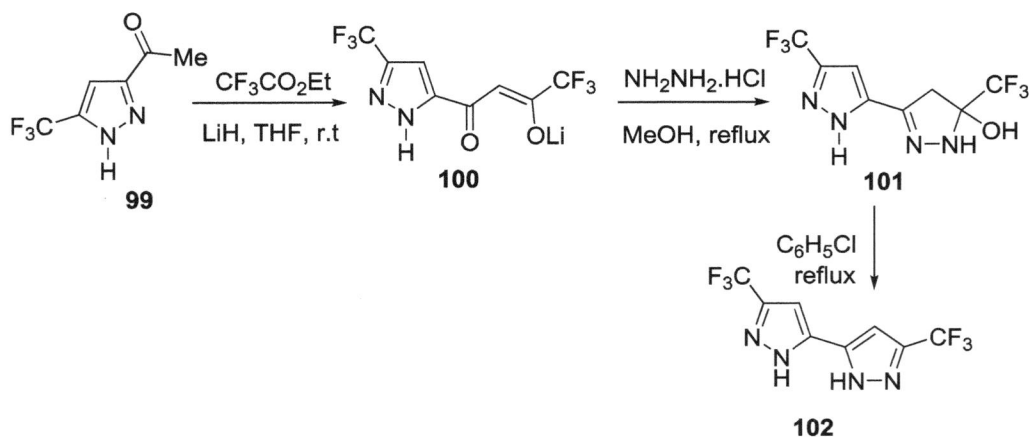

Scheme (28). Synthesis of the 3,3'-bipyrazoline derivatives **102**.

Cyclocondensation of 1,6-diethoxyhexa-1,5-diene-3,4-dione **103** with phenylhydrazine in *m*-cresol afforded 1,1'-diphenyl-3,3'-bipyrazole **104** in a high yield (Scheme **29**) [48].

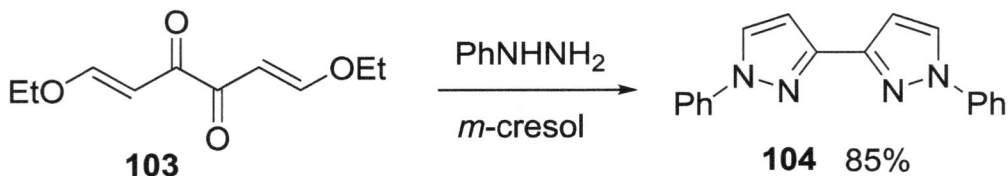

Scheme (29). Synthesis of the 3,3'-bipyrazoline derivatives **101**.

The unsubstituted 3,3'-bipyrazole **107** was prepared in 75% yield from the reaction of 1,6-diethoxy-1,5-hexadiene-3,4-dione (**105**) with hydrazine hydrate THF at room temperature, using *p*-toluenesulfonic acid as an acid catalyst. The reaction took place in a step-wise procedure with the elimination of water and ethanol through the intermediate **106** (Scheme **30**) [49].

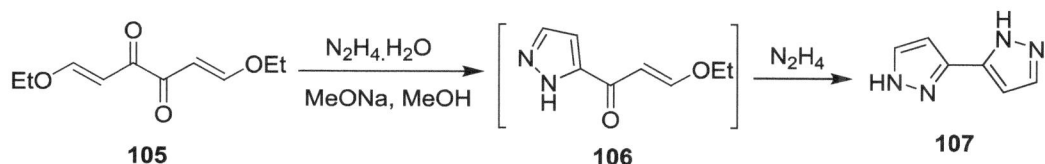

Scheme (30). Synthesis of the 3,3'-bipyrazoline derivatives **107**.

Heating 3-epoxypropionyl-2-pyrazolines **108** with hydrazine in refluxing methanol gave the 3,3'-bipyrazoline derivatives **109**. Acetylation of the latter compounds with acetyl chloride afforded the *N,N-bis*-acylated derivatives **110**. Hydrolysis of compound **110** (R = Ph) furnished 1-acetyl-4'-methyl-4-phen-l-4,5-dihydro-3,3'-bipyrazole **111**. Dehydrogenation of compound **110** (R = Ph) in the presence of elemental sulfur produced 4-methyl-4'-phenyl-3,3'-bipyrazole **112** (Scheme **31**) [50].

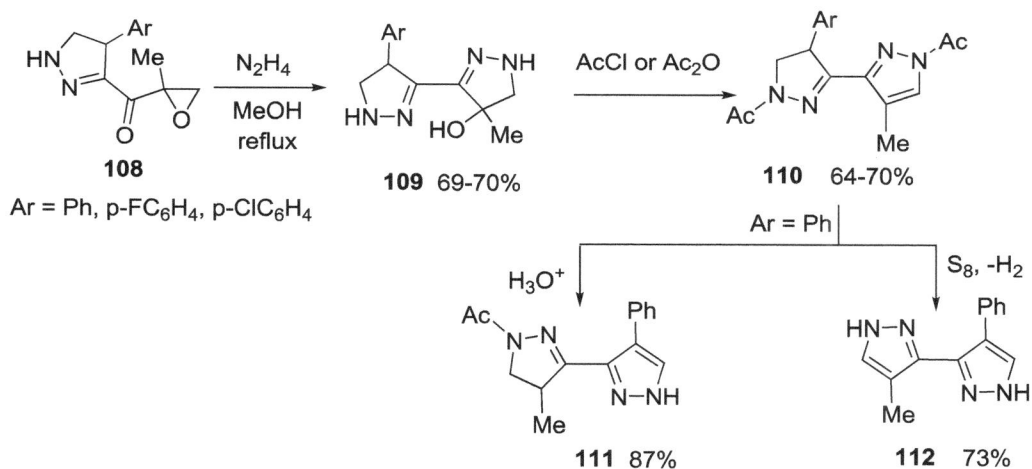

Scheme (31). Synthesis of the 3,3'-bipyrazoline derivatives **111** and **112**.

Condensation of 3-formylpyrazole derivative **113** with arylketones **114** afforded the α,β-unsaturated ketones **115**. The latter compounds **115** underwent cyclocondensation with hydrazines to afford the 3,3'-bipyrazoline derivatives **116** (Scheme **32**) [51].

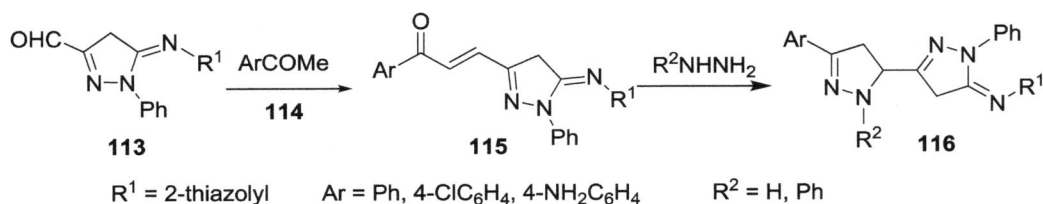

Scheme (32). Synthesis of the 3,3'-bipyrazoline derivatives **116**.

Heating a mixture of ethyl 3-acetyl-1-aryl-5-phenyl-1*H*-pyrazole-4-carboxylates **117** and the appropriate aromatic aldehydes in ethanolic sodium hydroxide solution furnished the corresponding pyrazolylchalcone derivatives **118** in high yields. The reaction of the latter chalcones **118** with hydrazine in ethanol afforded the bipyrazole derivatives **119** in high yields (Scheme **33**) [13]. The ^{1}H NMR spectral data of compound **119** (Ar = Ph, Ar` = 4-MeO-C$_6$H$_4$) was as following: δ = 2.90–3.0 (m, 1H, $H_c$), 3.42–3.56 (m, 1H, $H_b$), 3.74 (s, 3H, OCH$_3$), 4.50 (s,2H, NH$_2$), 4.78–4.95 (t, 1H, $H_a$), 6.90–7.35 (m, 14H, ArH`s), 7.85 (s, 1H, NH), 10.38 (s, 1H, NH) ppm.

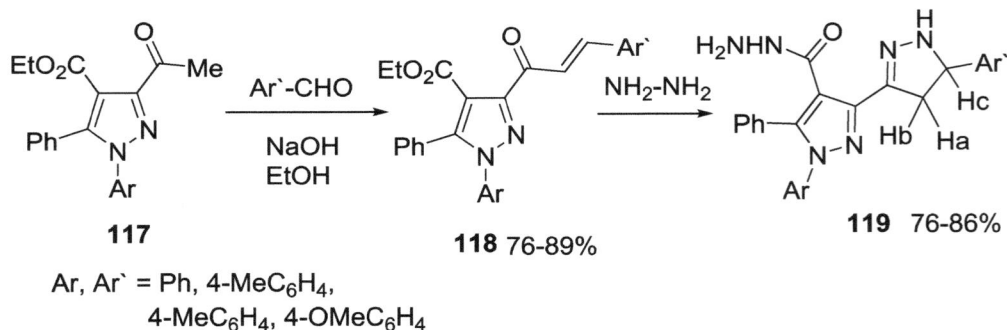

Ar, Ar` = Ph, 4-MeC$_6$H$_4$,
4-MeC$_6$H$_4$, 4-OMeC$_6$H$_4$

Scheme (33). Synthesis of the 3,3'-bipyrazole derivatives **98**.

Condensation of the acetylpyrazole derivative **120** with dimethylformamide-dimethylacetal (DMF-DMA) afforded the 3-[(*E*)-3-(*N,N*-dimethylamino) acryloyl]-1*H*-pyrazole-4-carboxylate derivative **121**. Treatment of the latter enaminone **121** with hydrazine hydrate in ethanol at reflux temperature afforded 3-(1*H*-pyrazol-3-yl)-1-(4-chlorophenyl)-5-phenyl-1*H*-pyrazole-4-carbohydrazide **122** in a high yield (Scheme **34**) [52, 53].

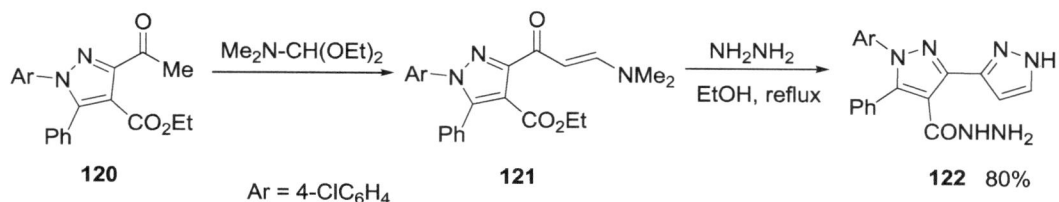

Scheme (34). Synthesis of the 3,3'-bipyrazole derivative **122**.

In a similar fashion, treatment of 3-acetyl-1H-pyrazole derivatives **123** with DMF-DMA afforded the corresponding enaminones **124** in very high yields. Treatment of the enaminones **124** with hydrazine hydrate in refluxing ethanol yielded the 3,3'-bipyrazoles **125** *via* initial addition of hydrazine to the enaminone double bond, followed by elimination of dimethylamine and water molecules to give **125** (Scheme **35**) [54].

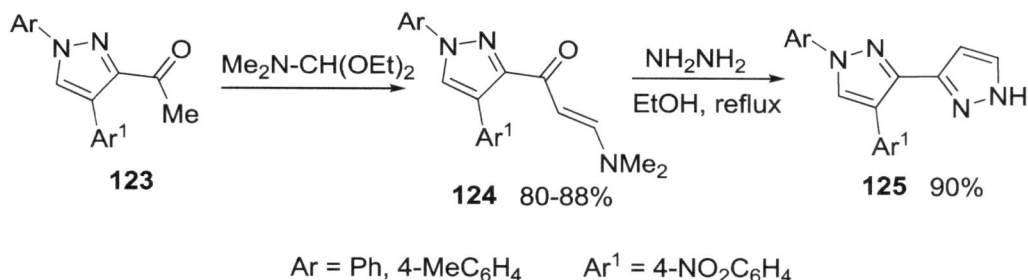

Scheme (35). Synthesis of the 3,3'-bipyrazole derivatives **125**.

When the 3-(pyrazol-3-yl)-3-oxo-propanenitrile derivative **126** was treated with dimethylformamide-dimethylacetal (DMF-DMA) in xylene at refluxing temperature, it afforded the enaminonitrile derivative **127** an excellent yield. The latter compounds **127** underwent further reaction with hydrazines and produced the 3,3'-bipyrazole derivatives **130** in good yields. The reaction proceeded through the mechanistic pathway depicted in Scheme (**36**) [55].

When the cyanoacetylpyrazole derivative **131** was treated with hydrazine hydrate in refluxing ethanol, it afforded the corresponding 3,3'-bipyrazole derivative **132** (Scheme **37**). In addition, the cyanoacetylpyrazole **131** reacted with phenyl isothiocyanate, using potassium hydroxide as a base, at room temperature, followed by the addition of methyl iodide to give the 3-(2-cyano-3-methylthio-3-phenylaminoacryloyl)pyrazole derivative **134**. Treatment of the latter **134** with hydrazine hydrate in refluxing ethanol gave the 3,3'-bipyrazole-4,4'-dicarbonitrile derivative **136** in a high yield. The reaction of the pyrazole derivative **131** with the pyrazolylhydrazonoyl bromide **137** in ethanolic sodium ethoxide solution afforded the 3,3`-bipyrazole derivatives **138** in good yields (Scheme 37) [25].

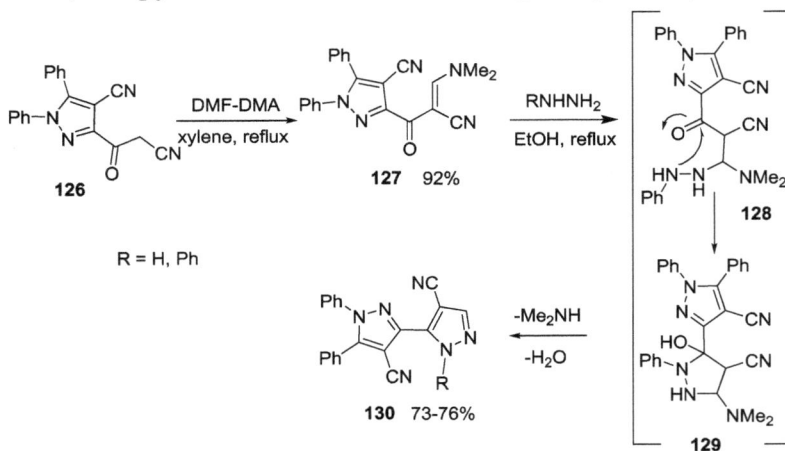

Scheme (36). Synthesis of the 3,3'-bipyrazole derivatives **130**.

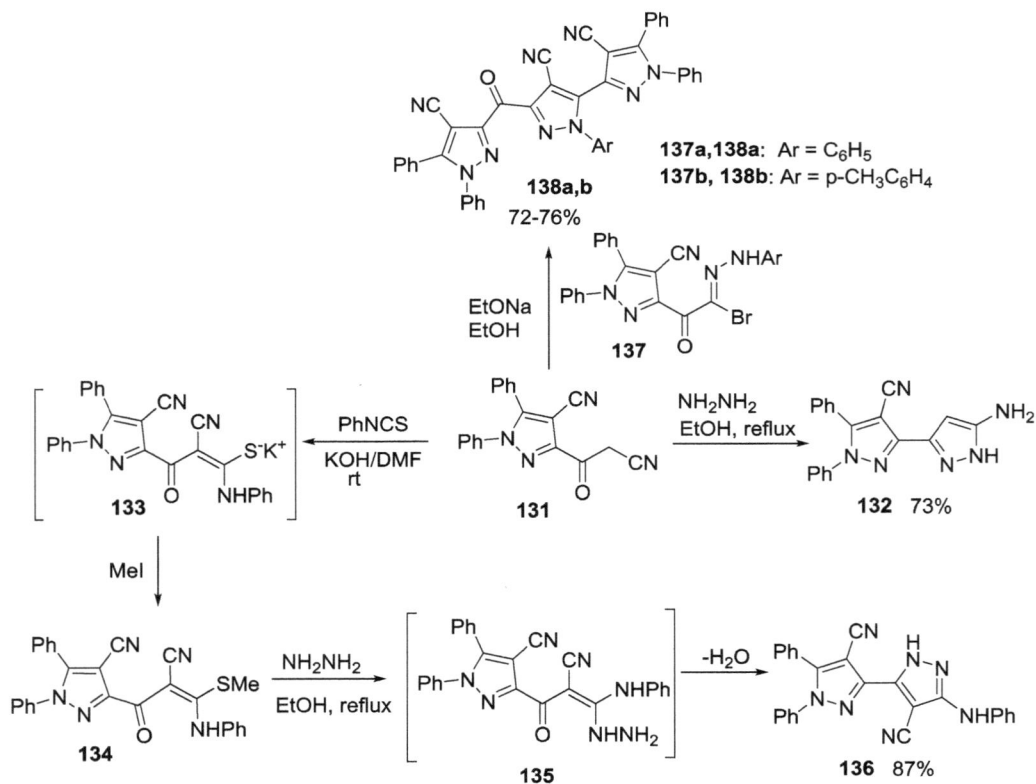

Scheme (37). Synthesis of the 3,3'-bipyrazole derivatives **136,138**.

The reaction of 3-bromoacetylpyrazole-4-carbonitrile **139** with sodium benzenesulfinate in refluxing ethanol produced the 3-(2-(phenylsulfonyl)acetyl)-pyrazole-4-carbonitrile derivative **140**. Treatment of the ketosulfone **140** with the hydrazonoyl chlorides **141** in an ethanolic solution of sodium ethoxide at ambient temperature yielded the corresponding 3,3`-bipyrazole derivatives **143** in good yields (Scheme **38**) [56].

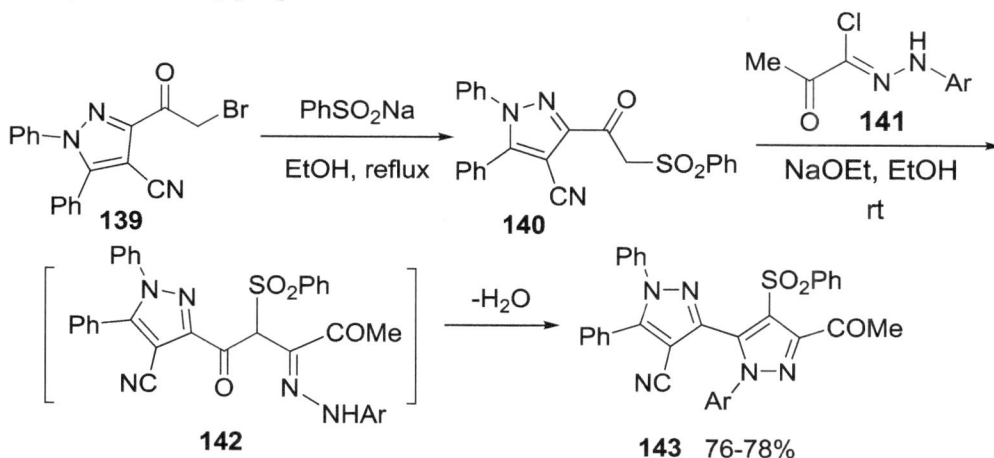

Scheme (38). Synthesis of the 3,3'-bipyrazole derivatives **143**.

Next, 3,3'-bi-isothiazole-4,4'-dicarbonitrile **144** was smoothly converted into the 3,3'-bipyrazole-4,4'-dicarbonitrile **147**, in a good yield, when was heated in anhydrous hydrazine under neat conditions according to the supposed mechanism outlined in Scheme (**39**) [57].

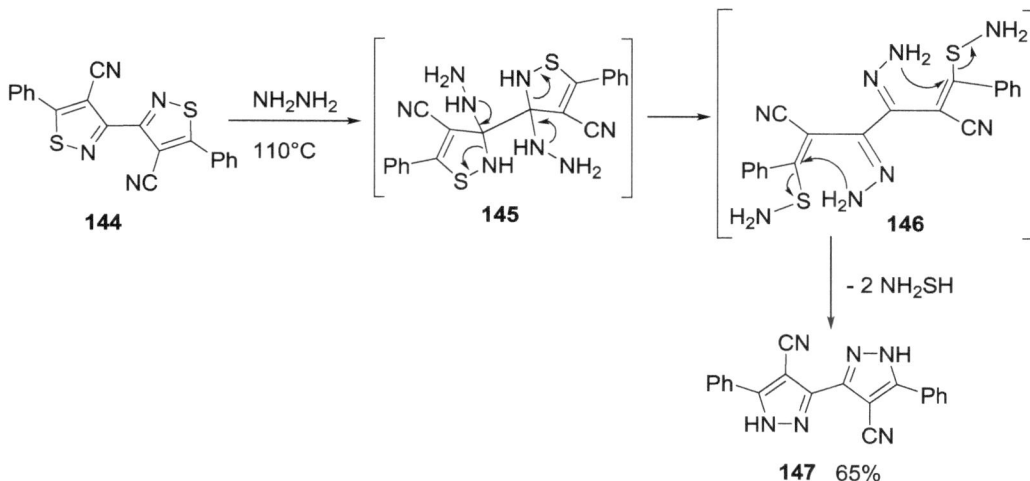

Scheme (39). Synthesis of the 3,3'-bipyrazole derivative **147**.

2.3. From Metal Catalyzed C-H Activation Reactions

Treatment of 1-benzyloxypyrazole **148** with butyllithium in THF solution followed by addition of zinc chloride in THF afforded 1-benzyloxypyrazol-5-ylzinc(II) chloride **149**. The latter compound underwent Negishi cross-coupling upon treatment with 5-iodo-1-benzyloxypyrazole **150** to afford 1,1'-(dibenzyloxy)-3,3'-bipyrazole **151**. The reaction of the latter 3,3'-bipyrazole **151** with an excess amount of iodine chloride in the presence of potassium carbonate as a base resulted in the formation of 4,4'-diiodo-3,3'-bipyrazole derivative **152** in an excellent yield. The Negishi reaction conditions were also applied to insert two phenyl groups at C-4 and C-4` in 3,3'-bipyrazole **152**. Therefore, the reaction took place by double iodine–magnesium exchange to the dimagnesium species, followed by exchange of magnesium with zinc, and finally, cross-coupling with iodobenzene, afforded 1,1'-(dibenzyloxy)-4,4'-diphenyl-3,3'-bipyrazole **154** in 70% yield. The conducted process above had a disadvantage where a huge amount of phenylmagnesium chloride was employed to obtain the intermediate **153**. Interestingly, Suzuki reaction of 4,4'-diiodo-3,3'-bipyrazole derivative **152** with phenylboronic acid proceeded efficiently to afford the 4,4'-diphenyl-3, 3'-bipyrazole **154** in a high yield. When compound **154** was heated in the presence of concentrated sulfuric acid debenzylation was occurred to afford the corresponding 1,1'-(dihydroxy)-4,4'-diphenyl-3,3'-bipyrazole **155** (Scheme **40**) [58, 59].

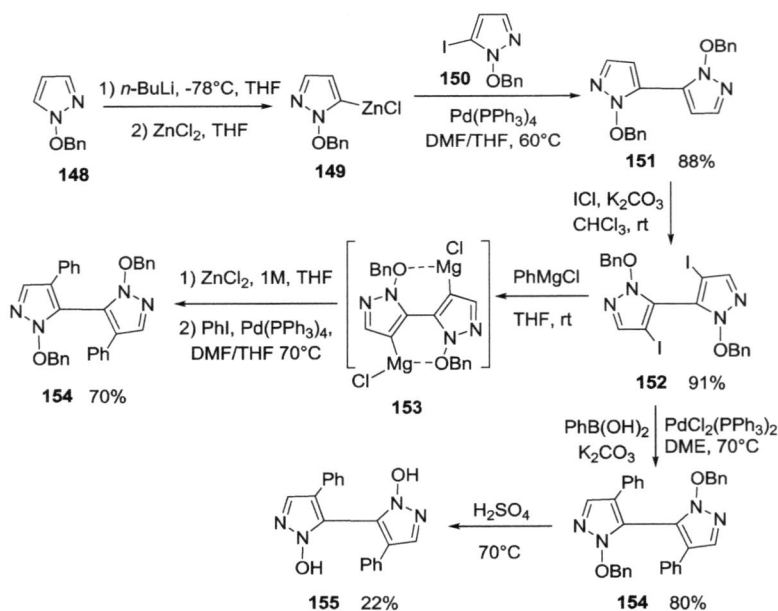

Scheme (40). Synthesis of the 3,3'-bipyrazole derivative **155**.

3,5-Dichloropyrazoles **156** reacted with Ni(cod)$_2$ complex [cod = 1,5-cyclooctadiene] using 2,2′-bipyridine as a ligand in dimethylformamide to afford the *bis*(pyrazolyl)nickel(II) complexes **157** in good yields. The reaction proceeded *via* oxidative addition where the C(5)–Cl bond of the two pyrazole units reacted with Ni(0)L$_m$. Treatment of complexes **157** with nitric acid at ambient temperature led to a reductive elimination to afford the 3,3′-dichloro-3,3′-bipyrazole derivatives **158** in good to excellent yields (Scheme **41**) [5, 60].

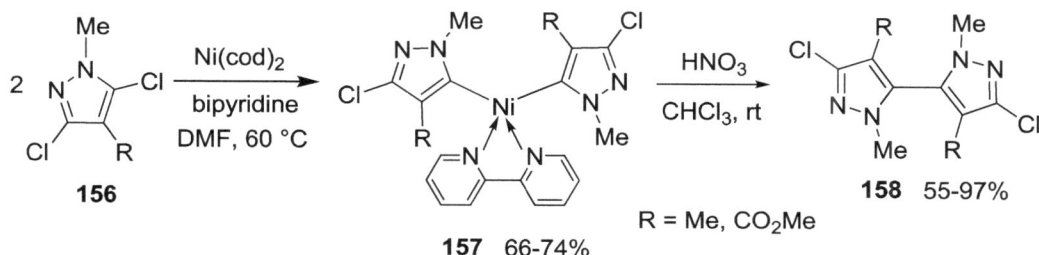

Scheme (41). Synthesis of the 3,3'-bipyrazole derivatives **158**.

3. REACTIONS OF 3,3`-BIPYRAZOLE DERIVATIVES

3.1. Nitration of 3,3`-Bipyrazole Derivatives

Nitration of **159** followed by oxidation of the product **160** gave the dinitro-bipyrazole-dicarboxylic acid **161**. Treatment of **161** with thionyl chloride followed by aqueous ammonia at room temperature resulted in the formation of bipyrazole-amide **163** in 62% yield, which was treated with bromine/NaOH to give the bipyrazole-amine **164** an 80% yield. The reaction of **164** with 2.2 equiv. of sodium nitrite in dilute sulfuric acid at 5°C, followed by the addition of sodium nitroacetonitrile led to the formation of the 4,4'-dinitro-5,5'-diamino-2H,2'H-(3,3'-bipyrazole) derivative **165** in 70% yield. Finally, 3,3',8,8'-tetranitro-[7,7'-bipyrazolo[5,1-c] [1, 2, 4]triazine]-4,4'-diamine (**166**) was prepared (in 88% yield) by cyclization of **165** in a mixture of methanol and water at reflux (Scheme **42**) [36, 61].

Reaction of 2H,2'H-3,3'-bipyrazole (**167**) with a mixture of acetic anhydride and 100% nitric acid gave the di(*N*-nitro)-pyrazole derivative **168,** which upon heating in benzonitrile at 140 °C rearranged to afford the di(*C*-nitro)-pyrazole derivative **169**. Nitration of **169** with a mixture of H$_2$SO$_4$ and 100% HNO$_3$ at 100 °C afforded 4,4',5,5'-tetranitro-2H,2'H-3,3'-bipyrazole (TNBP) (**170**). Further, nitration of **167** with a mixed acid (H$_2$SO$_4$ + 100% HNO$_3$) at 100 °C gave **171,** which was converted into 1,1',4,4'-tetranitro-1H,1'H-3,3'-bipyrazole (**172**) using trifluoroacetic anhydride and ammonium nitrate at 0 °C. In addition, the 2,2',5,5'-tetranitro-2*H*,2'*H*-3,3'-bipyrazole (**173**), as a structural isomer of **172**, was synthesized from reaction of **169** with acetic anhydride and nitric acid (Scheme **43**) [6, 49].

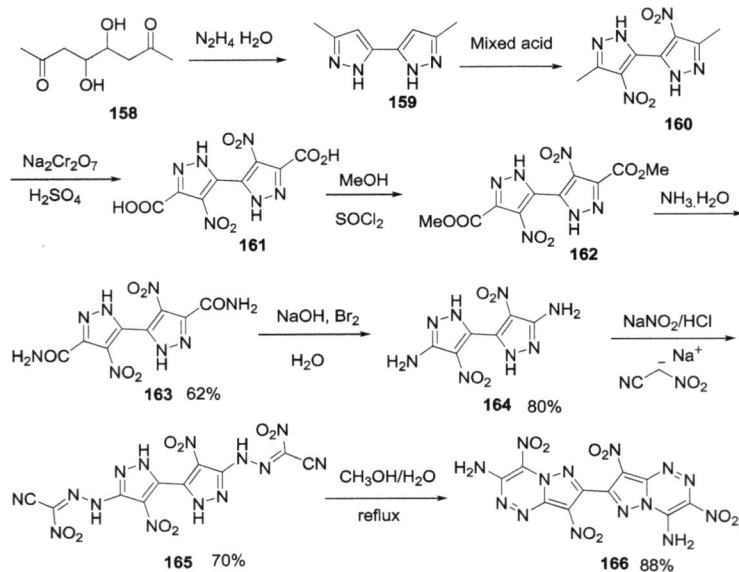

Scheme (42). Synthesis of the 3,3'-bipyrazole derivative **166**.

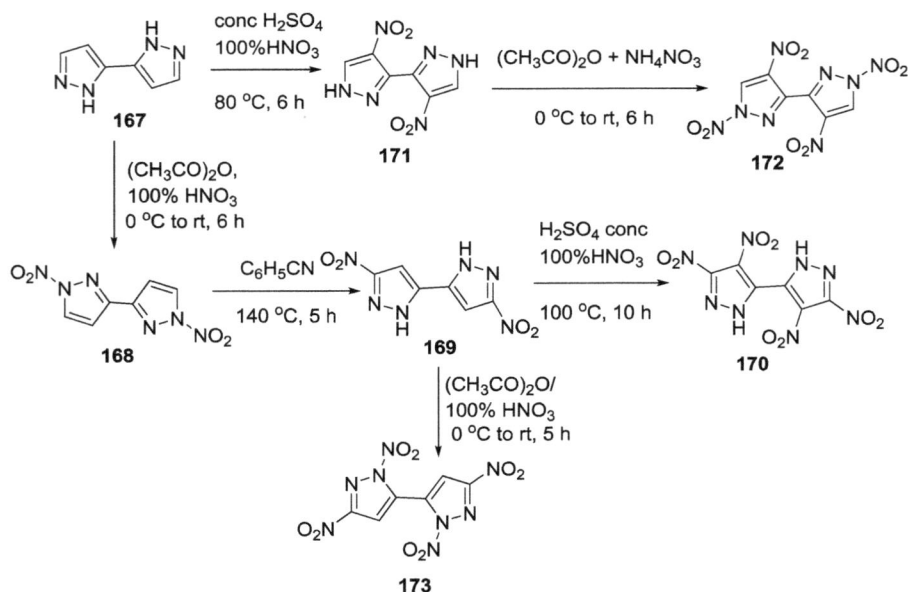

Scheme (43). Synthesis of 3,3'-bipyrazole derivatives **168-173**.

Alternatively, 4,4'-dinitro-1*H*,1'*H*-[3,3'-bipyrazole]-5,5'-diamine (**176**) was prepared from *N,N*-dinitrobipyrazole derivative **172** as depicted in Scheme (**44**). The *N,N*-dinitrobipyrazole derivative **172** underwent a *cine*-substitution reaction in the presence of anionic carbon and nitrogen nucleophiles under mild reaction conditions to yield the 5,5'-disubstituted derivatives **174** and **177** (Scheme **44**). Acidic hydrolysis of the cyano functions of bipyrazole **174** at 50-60 °C gave the amide **175**. Treatment of **177** with triphenylphosphine at room temperature furnished the diphosphiminopyrazole **178**. Hoffmann rearrangement of either the amide **175** or the diphosphimine **178** led to the formation of the corresponding diamine **176** in an acceptable yield [62]. The ^{1}H NMR of **176** exhibited peaks at δ 7.31 (s, 2H, NH$_2$), 12.49 (br. s, 1H, NH pyrazole); its ^{13}C NMR showed peaks at δ 147.7 (C-3), 138.9 (C-5), 116.8 (C-4), and its ^{14}N NMR showed a peak at δ -18.10 due to NO$_2$ group.

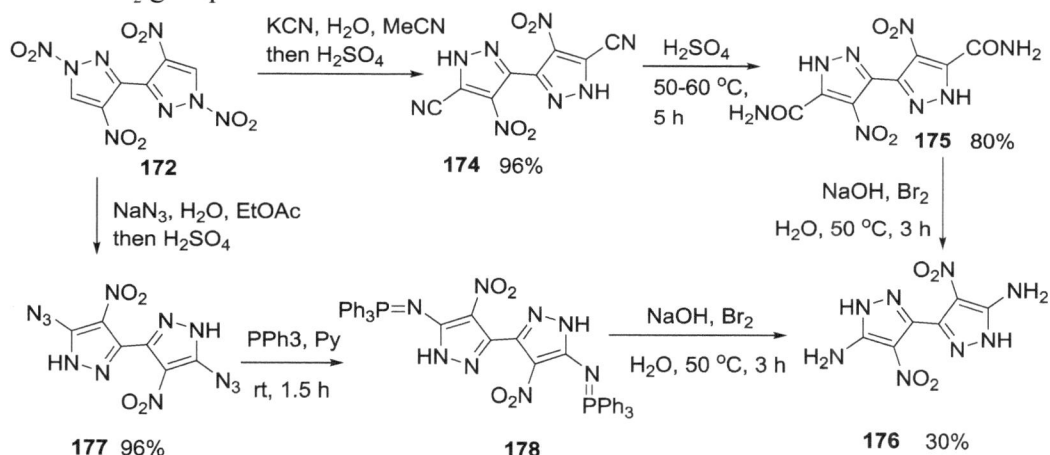

Scheme (44). Alternative synthesis of 3,3'-bipyrazole derivative **176**.

N-Amination reaction of 4,4`,5,5`-tetranitro-2*H*,2*H*-3,3-bipyrazole (**170**) was performed using O-*p*-toluenesulfonylhydroxylamine (THA) to give the di-*N*-amino derivative **179** in 66% yield. Reaction of **179** with *tert*-butyl hypochlorite (*t*BuOCl) afforded the fused-ring system **180** (Scheme **45**) [7].

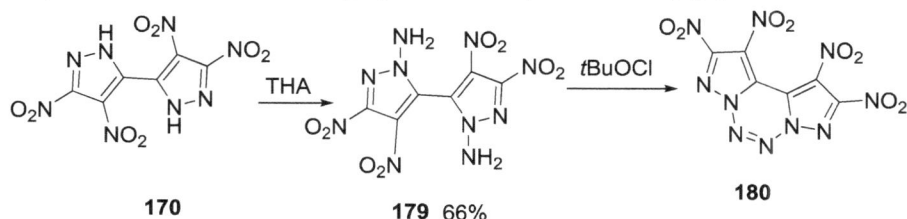

Scheme (45). Synthesis of the 3,3'-bipyrazoline derivatives **179** and **180**.

Furthermore, the 4,4`,5,5`-tetranitro-2`-trinitromethyl-3,3`-bipyrazole derivative **182** was obtained in a low yield by alkylation of **170** with bromoacetone followed by nitration using a mixture of concentrated sulfuric acid and 100% nitric acid for seven days. The reaction of compound **170** with O-*p*-toluenesulfonylhydroxylamine (THA) in the presence of DBU afforded the *bis*-1,1`-diamino-3,3`-bipyrazole derivative **179**. Nitration of the latter 3,3`-bipyrazole **179** with nitronium tetrafluoroborate (NO$_2$BF$_4$) using potassium acetate as a base led to the formation of the monopotassium salt **183**. Treatment of compound **170** with potassium bicarbonate gave the dipotassium salt **184** (Scheme **46**) [8].

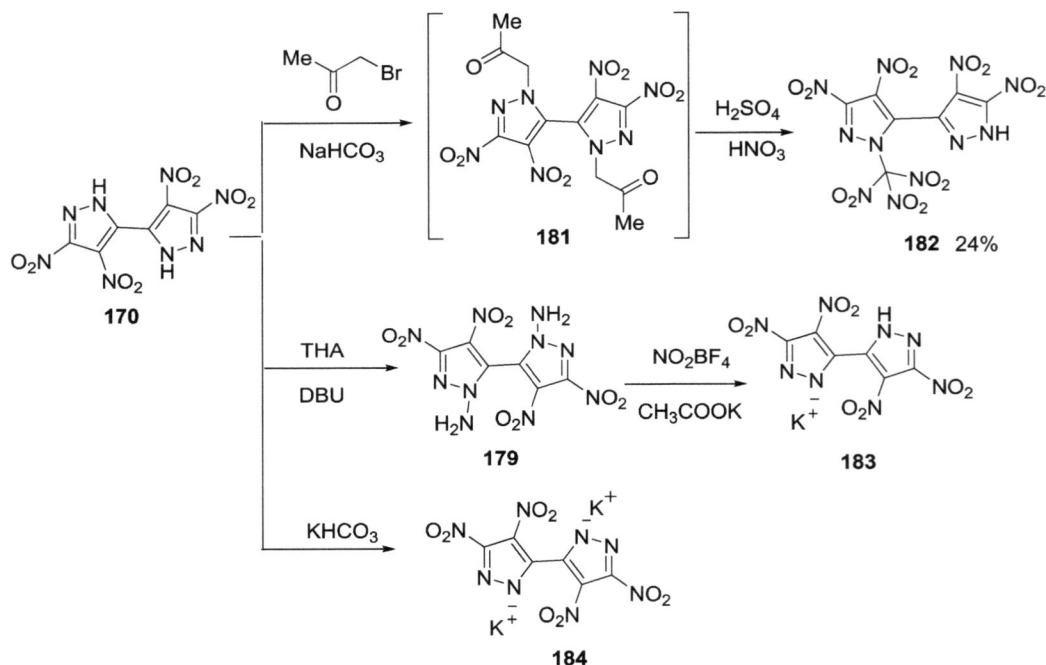

Scheme (46). Synthesis of the 3,3'-bipyrazole derivatives **182-184**.

1,1'-Difluoroamino-3,3',4,4'-tetranitro-3,3'-bipyrazole **186** was prepared in a low yield *via* treatment of 3,3',4,4'-tetranitro-3,3'-bipyrazole **170** with O-fluorosulfonyl-*N,N*-difluorohydroxylamine **185** under phase-transfer catalysis condition using PEG-400 and NaHCO$_3$. However, reaction of **170** with NaF/NaOH in methanol followed by addition of F$_2$/N$_2$ at -70 °C afforded 1,1'-difluoro-3,3',4,4'-tetranitro-3,3'-bipyrazole **187** (Scheme **47**) [63].

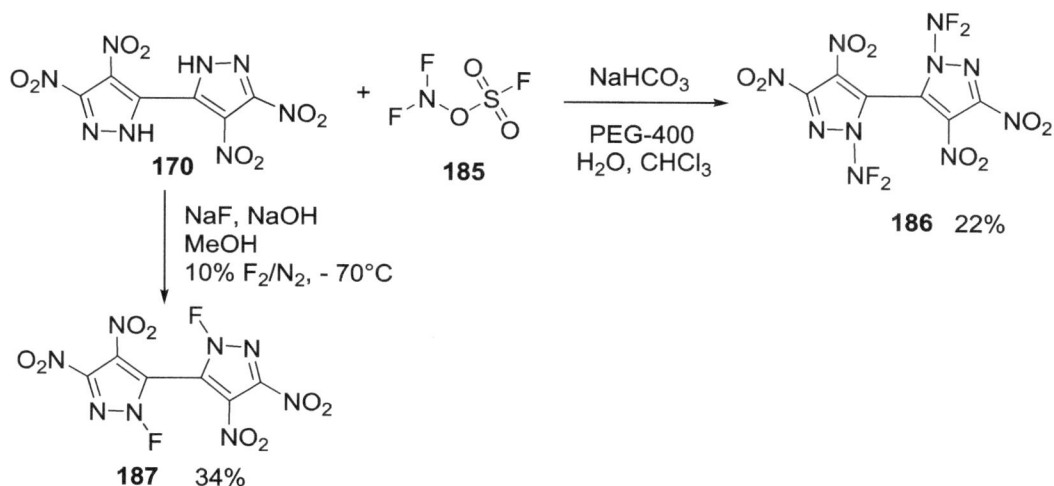

Scheme (47). Synthesis of the 3,3'-bipyrazole derivatives **186-187**.

3.2. Miscellaneous Reactions

3,3',4,4',5,5'-hexahydro-3,3'-bipyrazole **188** underwent oxidative dehydrogenation when was treated with MnO_2 in benzene at room temperature and afforded a mixture of 3,3'-bipyrazole **189** and 3-cyclopropyl-1*H*-pyrazole **190**. Formation of the bipyrazole structure **189** was due to oxidation followed by tautomerism, and formation of **190** was due to extrusion of nitrogen molecule (Scheme **48**) [64, 65].

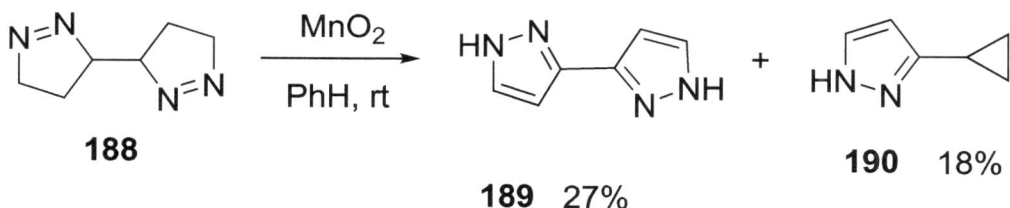

Scheme (48). Synthesis of the 3,3'-bipyrazole derivatives **189**.

Palladium(II) and platinum(II) complexes of 5,5'-dimethyl-3,3'-bipyrazole **191** were reported to have potential anti-tumor properties [66].

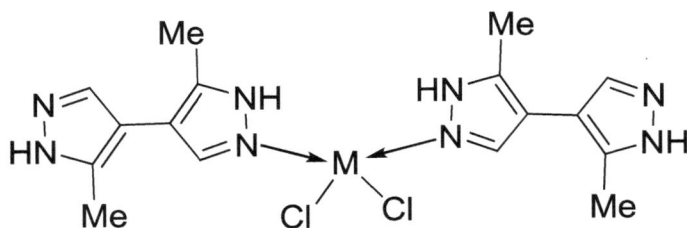

191   M = Pd, Pt

Treatment of the bidentate 3,3'-bipyrazole ligands **192** and **189** with the monohydrido ruthenium(II) complex **193** gave the corresponding carbonyl(hydrido)*bis*-(triphenylphosphane)ruthenium(II) complexes **194** in 65-78% yields. The ruthenium(II) complexes **194** showed catalytic activity and transfer of hydrogen in catalyzed hydrogenation reactions (Scheme **49**) [2, 3].

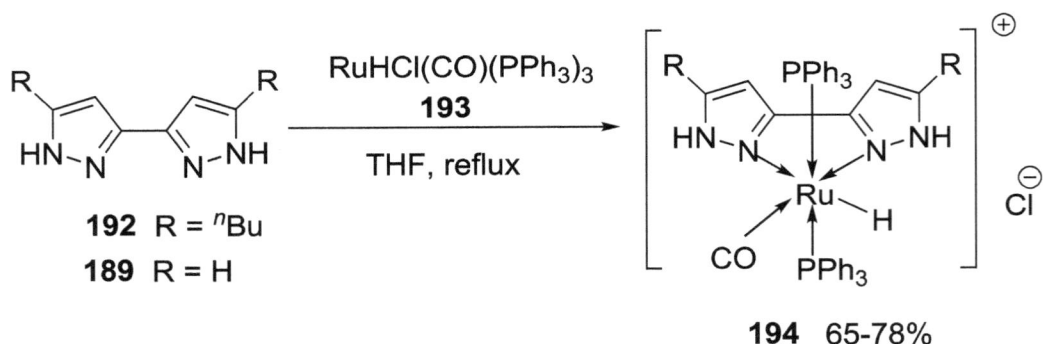

192   R = ^{n}Bu
189   R = H

194   65-78%

Scheme (49). Synthesis of the 3,3'-bipyrazole derivatives **186-187**.

CONCLUSIONS

In this chapter, the structurally related 3,3`-bipyrazoles, 3,3`-bipyrazolines of 3-(pyrazol-3-yl)pyrazolines were synthesized and their chemical reactions were also reported. A wide range of synthetic pathways was utilized to fulfill the construction of the presented 3,3`-bipyrazole systems, including cyclocondensation of tetracarbonyl or dihydroxydicarbonyl compounds or pyrazoles having a difunctional group with hydrazines, in addition to 1,3-dipolar cycloaddition of pyrazolyl-nitrilimines or bis-nitrilimines with the appropriate olefins or acetylenes. Furthermore, metal catalysed C-H activation reactions of halopyrazoles with pyrazoleboronic acid derivatives led to the formation of 3,3`-bipyrazoles. The potential academic and industrial applications of 3,3`-bipyrazole derivatives and their metal complexes are driving forces for more scientific research on such classes of biheterocycles. The nitrated-3,3`-bipyrazoles were reported as metal-free primary explosives with high energetic properties and excellent thermal stability.

REFERENCES

[1]   El Ati, R.; Takfaoui, A.; El Kodadi, M.; Touzani, R.; Yousfi, E.B.; Almalki, F.A.; Hadda, T.B. Catechol oxidase and Copper (I/II) Complexes Derived from Bipyrazol Ligand: Synthesis, Molecular Structure Investigation of New Biomimetic Functional Model and Mechanistic Study. *Mater. Today Proc.*, **2019**, *13*, 1229-1237.
[http://dx.doi.org/10.1016/j.matpr.2019.04.092]

[2]   Jozak, T.; Zabel, D.; Schubert, A.; Sun, Y.; Thiel, W.R. Ruthenium Complexes Bearing N–H Acidic Pyrazole Ligands. *Eur. J. Inorg. Chem.*, **2010**, *2010*(32), 5135-5145.
[http://dx.doi.org/10.1002/ejic.201000802]

[3]   Romero, A.; Vegas, A.; Santos, A.; Cuadro, A.M. Reactivity of ruthenium (II) complexes with 1-hydroxymethyl-or 1-benzyl-3,5-dimethylpyrazole and similar functionalized bipyrazoles. X-Ray crystal structure of carbonylchlorohydrido (3, 5-dimethylpyrazole-N 2)-bis (triphenylphosphine) ruthenium (II). *J. Chem. Soc., Dalton Trans.*, **1987**, (1), 183-186.
[http://dx.doi.org/10.1039/dt9870000183]

[4]   Al-Fulaij, O.A.; Elassar, A.Z.A.; Dawood, K.M. Synthesis and characterization of new 3, 3-bipyrazole-4, 4-dicarboxylic acid derivatives and some of their palladium (II) complexes as pre-catalyst for Suzuki coupling reaction in water. *Eur. J. Chem.*, **2019**, *10*(4), 367-375.
[http://dx.doi.org/10.5155/eurjchem.10.4.367-375.1915]

[5]   Murakami, Y.; Yamamoto, T. Ni-Promoted Syntheses of New 3,3′-Dichloro-5,5′-bipyrazoles and Poly (bipyrazole-5, 5′-diyl) s and Isolation of Nickel Complexes Relevant to the Syntheses. *Bull. Chem. Soc. Jpn.*, **1999**, *72*(7), 1629-1635.
[http://dx.doi.org/10.1246/bcsj.72.1629]

[6]   Kumar, D.; Tang, Y.; He, C.; Imler, G.H.; Parrish, D.A.; Shreeve, J.M. Multipurpose Energetic Materials by Shuffling Nitro Groups on a 3,3′-Bipyrazole Moiety. *Chemistry*, **2018**, *24*(65), 17220-17224.
[http://dx.doi.org/10.1002/chem.201804418] [PMID: 30231192]

[7]   Tang, Y.; Kumar, D.; Shreeve, J.M. Balancing excellent performance and high thermal stability in a dinitropyrazole fused 1,2,3,4-tetrazine. *J. Am. Chem. Soc.*, **2017**, *139*(39), 13684-13687.
[http://dx.doi.org/10.1021/jacs.7b08789] [PMID: 28910088]

[8]   Tang, Y.; He, C.; Imler, G.H.; Parrish, D.A.; Jean'ne, M.S. Energetic derivatives of 4,4′,5,5′-tetranitr--2H,2′H-3,3′-bipyrazole (TNBP): synthesis, characterization and promising properties. *J. Mater. Chem. A Mater. Energy Sustain.*, **2018**, *6*(12), 5136-5142.
[http://dx.doi.org/10.1039/C7TA11172J]

[9]   Sarkar, A.; Mandal, T.K.; Rana, D.K.; Dhar, S.; Chall, S.; Bhattacharya, S.C. Tuning the photophysics of a bio-active molecular probe '3-pyrazolyl-2-pyrazoline'derivative in different solvents: dual effect of polarity and hydrogen bonding. *J. Lumin.*, **2010**, *130*(11), 2271-2276.
[http://dx.doi.org/10.1016/j.jlumin.2010.07.004]

[10]   Hsu, C.W.; Ly, K.T.; Lee, W.K.; Wu, C.C.; Wu, L.C.; Lee, J.J.; Lin, T.C.; Liu, S.H.; Chou, P.T.; Lee, G.H.; Chi, Y. Triboluminescence and metal phosphor for organic light-emitting diodes: functional Pt (II) complexes with both 2-pyridylimidazol-2-ylidene and bipyrazolate chelates. *ACS Appl. Mater. Interfaces,* **2016**, *8*(49), 33888-33898.
[http://dx.doi.org/10.1021/acsami.6b12707] [PMID: 27960361]

[11]   Liao, J.L.; Chi, Y.; Yeh, C.C.; Kao, H.C.; Chang, C.H.; Fox, M.A.; Low, P.J.; Lee, G.H. Near

infrared-emitting tris-bidentate Os-(II) phosphors: control of excited state characteristics and fabrication of OLEDs. *J. Mater. Chem. C Mater. Opt. Electron. Devices*, **2015**, *3*(19), 4910-4920.
[http://dx.doi.org/10.1039/C5TC00204D]

[12]   Chi, Y.; Yeh, H-H. Luminescent platinum(ii) complexes with bizolate chelates, U.S. Patent **2015**. US 9040702, B1.

[13]   Masaret, G.S. Synthesis, Structure Elucidation, and Biological Activities of Pyrazoles Against Human Lung and Hepatocellular Cancer. *J. Heterocycl. Chem.*, **2018**, *55*(9), 2123-2129.
[http://dx.doi.org/10.1002/jhet.3257]

[14]   Farag, A.M.; Shawali, A.S.; Abed, N.M.; Dawood, K.M. 1,3-Dipolar cycloaddition syntheses of 3,3'-bi(2-pyrazolines),3,3'-bipyrazoles and 3,3'-bi(1,2,4-triazoles). *Gazz. Chim. Ital.*, **1993**, *123*(8), 467-470.

[15]   Farag, A.M.; Kheder, N.A.; Budesinsky, M. Regioselective synthesis of polysubstituted 3,3'-bi--H-pyrazole derivatives *via* 1,3-dipolar cycloaddition reactions. *Tetrahedron*, **1997**, *53*(27), 9293-9300.
[http://dx.doi.org/10.1016/S0040-4020(97)00583-8]

[16]   Dawood, K.M.; Elwan, N.M. Synthesis of 3,3'-bipyrazole,5,5'-bi-1,3,4-thiadiazole and fused azole systems *via* bishydrazonoyl chlorides. *J. Chem. Res.*, **2004**, *2004*(4), 264-266.
[http://dx.doi.org/10.3184/0308234041209121]

[17]   Dawood, K.M.; Elassar, A.Z.A.; Al-Fulaij, O.A. Facile access to some new 3, 3'-bipyrazole-ester derivatives utilizing bis-hydrazonoyl chlorides. *J. Heterocycl. Chem.*, **2020**, *57*(1), 370-376.
[http://dx.doi.org/10.1002/jhet.3787]

[18]   Dawood, K.M. Regio-and stereoselective synthesis of bis-spiropyrazoline-5,3'-chroman (thiochroman)-4-one derivatives *via* bis-nitrilimines. *Tetrahedron*, **2005**, *61*(22), 5229-5233.
[http://dx.doi.org/10.1016/j.tet.2005.03.083]

[19]   Behbehani, H.; Ibrahim, H.M.; Dawood, K.M. Ultrasound-Assisted Regio- and Stereoselective Synthesis of Bis-[1`,4'-diaryl-1-oxo-spiro-benzosuberane-2,5`-pyrazoline] Derivatives *via* 1,3-Dipolar Cycloaddition. *RSC Advances*, **2015**, *5*(33), 25642-25649.
[http://dx.doi.org/10.1039/C5RA02972D]

[20]   Behbehani, H.; Dawood, K.M.; Ibrahim, H.M.; Mostafa, N.S. Regio-and stereoselective route to bis-[3-methyl-1,1',4'-triaryl-5-oxo-spiro-pyrazoline-4,5'-pyrazoline] derivatives *via* 1,3-dipolar cycloaddition under sonication. *Arab. J. Chem.*, **2018**, *11*(7), 1053-1060.
[http://dx.doi.org/10.1016/j.arabjc.2017.05.016]

[21]   Arrieta, A.; Carrillo, J.R.; Cossio, F.P.; Diaz-Ortiz, A.; Gómez-Escalonilla, M.J.; de la Hoz, A.; Moreno, A.; Langa, F. Efficient tautomerization hydrazone-azomethine imine under microwave irradiation. Synthesis of [4,3'] and [5,3']bipyrazoles. *Tetrahedron*, **1998**, *54*(43), 13167-13180.
[http://dx.doi.org/10.1016/S0040-4020(98)00798-4]

[22]   Carrillo, J.R.; Cossio, F.P.; Diaz-Ortiz, A.; Gomez-Escalonilla, M.J.; de la Hoz, A.; Moreno, A.; Lecea, B.; Moreno, A.; Prieto, P. A complete model for the prediction of 1H-and 13C-NMR chemical shifts and torsional angles in phenyl-substituted pyrazoles. *Tetrahedron*, **2001**, *57*(19), 4179-4187.
[http://dx.doi.org/10.1016/S0040-4020(01)00291-5]

[23]   Mukherjee, A.; Mahalanabis, K.K. Preparation of 4-Cyano-3-methyl-1-phenyl-1H-pyra-ole-5-(4-bromophenyl) nitrile Imine: Regio-and Stereoselective Synthesis of a New Class of Substituted 3-Pyrazolines and Pyrazoles. *Heterocycles*, **2009**, *78*(4), 911.
[http://dx.doi.org/10.3987/COM-08-11531]

[24]   Hassaneen, H.M.; Shawali, A.S.; Elwan, N.M. A convenient synthesis of 3, 5'-bipyrazolyl derivatives *via* hydrazonyl halides. *Heterocycles*, **1990**, *31*(6), 10411047.
[http://dx.doi.org/10.3987/COM-90-5310]

[25]   Dawood, K.M.; Ragab, E.A.; Farag, A.M. Synthesis of bipyrazole and 1,3,4-thiadiazole derivatives. *J. Chem. Res.*, **2009**, *2009*(10), 630-634.

[http://dx.doi.org/10.3184/030823409X12528547964044]

[26]   Padwa, A.; Meske, M.; Rodriguez, A. 1,3-Dipolar cycloaddition chemistry of 2, 3-bis (phenylsulfonyl)-1, 3-Diene with Diazoalkanes. *Heterocycles,* **1995**, *40*(1), 191-204.
[http://dx.doi.org/10.3987/COM-94-S8]

[27]   Franck-Neumann, M.; Geoffroy, P.; Lohmann, J.J. Additions 1, 3-dipolaires du diazo-2-propane sur le diacetylene et evolution photochimique des 3H-pyrazoles formes. *Tetrahedron Lett.,* **1983**, *24*(17), 1775-1778.
[http://dx.doi.org/10.1016/S0040-4039(00)81767-0]

[28]   Mahendran, V.; Pasumpon, K.; Shanmugam, S. An Easy Access to Bipyrazoles and Unusual Demethylation of Methyl Phosphorous Ester: Exploring the Synthetic Utility of Bestmann-Ohira Reagent. *ChemistrySelect,* **2017**, *2*(9), 2866-2869.
[http://dx.doi.org/10.1002/slct.201700239]

[29]   Terashima, S.; Newton, G.N.; Shiga, T.; Oshio, H. Planar trinuclear complexes with linear arrays of metal ions. *Inorg. Chem. Front.,* **2015**, *2*(2), 125-128.
[http://dx.doi.org/10.1039/C4QI00172A]

[30]   Bouabdallah, I.; Ramdani, A.; Zidane, I.; Touzani, R.; Eddike, D.; Radi, S.; Haidoux, A. Regioselective synthesis and crystal structure of 1,1'-dibenzyl-5,5'-diisopropyl-3,3'-bipyrazole. *J. Marocain. Chim. Heterocycl.,* **2004**, *3*(1), 39-44.

[31]   Bouabdallah, I.; Zidane, I.; Touzani, R.; Ramdani, A.; Jalbout, A.F.; Trzaskowski, B. New 1-(ethy--ethanoate-yl)-5,5'-diisopropyl-3,3'-bipyrazole. *Molbank,* **2006**, *M491*(5), 1-3.
[http://dx.doi.org/10.3390/M491]

[32]   Bouabdallah, I.; Ramdani, A.; Zidane, I.; Touzani, R. 1,1'-dibenzyl-5,5'-diphenyl-3,3'-bipyrazole. *Molbank,* **2006**, *M482*(4), 1-2.
[http://dx.doi.org/10.3390/M482]

[33]   Bouabdallah, I.; Zidane, I.; Hacht, B.; Ramdani, A.; Touzani, R. Liquid-liquid extraction of metals by using new bipyrazolic compounds. *J. Mater. Environ. Sci.,* **2010**, *1*, 20-24.

[34]   Bouabdallah, I.; Ramdani, A.; Zidane, I.; Touzani, R.; Eddike, D.; Haidoux, A. Synthesis and crystal structure of ac, c-linked bipyrazole compound: 1,1'-bis (4-nitrophenyl)-5, 5'-diisopropyl-3,-'-bipyrazole. *J. Marocain Chim. Heterocycl.,* **2006**, *5*(1), 52-75.

[35]   Bouabdallah, I.; Ramdani, A.; Zidane, I.; Eddike, D.; Tillard, M.; Belin, C. E, **2005**, *61*(12), o4243-o4245.

[36]   Shironina, T.M.; Igidov, N.M.; Kuz'minykh, E.N.; Kon'shina, L.O.; Kasatkina, Yu.S.; Kuz'minykh, V.O. 1,3,4,6-Tetracarbonyl compounds: IV. Reaction of 3, 4-dihydroxy-2,4-hexadiene-1,6-diones with hydrazine and arylhydrazines. *Russ. J. Org. Chem.,* **2001**, *37*(10), 1486-1494.
[http://dx.doi.org/10.1023/A:1013431407273]

[37]   Kovac, S.; Rapic, V.; Lacan, M. Synthese und einige reactionen des 1,6-Bis (p-hydroxyphenyl--1,3,4,6-hexantetrons. *Liebigs Ann. Chem.,* **1984**, 1755-1758.
[http://dx.doi.org/10.1002/jlac.198419841016]

[38]   Kozminykh, V.O.; Konshina, L.O.; Igidov, N.M. 1,3,4,6-Tetracarbonyl Compounds. I. The Novel Synthesis of 1,6-Diaryl-3,4-dihydroxy-hexa-2,4-diene-1,6-diones from 5-Aryl-furan-2,3-diones. *J. Prakt. Chem.,* **1993**, *335*(8), 714-716.
[http://dx.doi.org/10.1002/prac.19933350813]

[39]   Bouabdallah, I.; Zidane, I.; Touzani, R.; Malek, F.; Ramdani, A.; Jalbout, A.F.; Trzaskowski, B. Synthesis of new C,C-linked bipyrazole and comparative theoretical calculations. *Chem. Sci. Trans.,* **2014**, *3*(2), 805-811.

[40]   Barrera, J.; Smolenski, V.A.; Jasinski, J.P.; Pastrán, J. *Synthesis and Crystal Structure of C1-Symmetric 3,3'-Bi(1,1'-dinaphthyl-camphopyrazole)*; J. Crystall, **2016**, p. 1217867.

[41]   El-Massaoudi, M.; Radi, S.; Mabkhot, Y.N.; Al–Showiman, S.S.; Ghabbour, H.A.; Ferbinteanu, M.; Adarsh, N.N.; Garcia, Y. Cu (II) and Mn (II) coordination complexes constructed by C linked bispyrazoles: Effect of anions and hydrogen bonding on the self assembly process. *Inorg. Chim. Acta,* **2018**, *482*, 411-419.
[http://dx.doi.org/10.1016/j.ica.2018.06.041]

[42]   Radi, S.; Tighadouini, S.; Hadda, T.B.; Akkurt, M.; Özdemir, N.; Sirajuddin, M.; Mabkhot, Y.N. Crystal structure of (Z)-1-(1,5-dimethyl-1H-pyrazol-3-yl)-3-hydroxybut-2-en-1-one. *Zeitsch. Krist.-. New Cryst. Struct,* **2016**, *231*(2), 617-618.

[43]   Smith, C.D.; Tchabanenko, K.; Adlington, R.M.; Baldwin, J.E. Synthesis of linked heterocycles *via* use of bis-acetylenic compounds. *Tetrahedron Lett.,* **2006**, *47*(19), 3209-3212.
[http://dx.doi.org/10.1016/j.tetlet.2006.03.052]

[44]   Kuz'minykh, V.O.; Andreichikov, Y.S. Recyclization of 2-alkoxycarbonylmethyl-1,5-diar-l-2-hydroxy-2,3-dihydropyrrol-3-ones by hydrazine. *Chem. Heterocycl. Compd.,* **1988**, *24*(12), 1406-1406.
[http://dx.doi.org/10.1007/BF00486691]

[45]   Kuz'minykh, V.O.; Igidov, N.M.; Andreichikov, Y.S. Chemistry of 2-methylene-2, 3-dihydro-3-pyranones. *Chem. Heterocycl. Compd.,* **1992**, *28*(8), 861-867.
[http://dx.doi.org/10.1007/BF00531315]

[46]   Bazhin, D.N.; Chizhov, D.L.; Röschenthaler, G.V.; Kudyakova, Y.S.; Burgart, Y.V.; Slepukhin, P.A.; Saloutin, V.I.; Charushin, V.N. A concise approach to CF_3-containing furan-3-ones,(bis) pyrazoles from novel fluorinated building blocks based on 2,3-butanedione. *Tetrahedron Lett.,* **2014**, *55*(42), 5714-5717.
[http://dx.doi.org/10.1016/j.tetlet.2014.08.046]

[47]   Yeh, H.H.; Ho, S.T.; Chi, Y.; Clifford, J.N.; Palomares, E.; Liu, S.H.; Chou, P.T. Ru (II) sensitizers bearing dianionic biazolate ancillaries: ligand synergy for high performance dye sensitized solar cells. *J. Mater. Chem. A Mater. Energy Sustain.,* **2013**, *1*(26), 7681-7689.
[http://dx.doi.org/10.1039/c3ta10988g]

[48]   Mitsury, U.; Kawaharasaki, M. *Makromol. Chem.,* **1981**, *192*, 837.

[49]   Vicente, V.; Fruchier, A.; Elguero, J. The Effenberger's synthesis of 3,3′-bipyrazole revisited. *ARKIVOC,* **2004**, (iii), 5-10.

[50]   Zvonok, A.M.; Kuz'menok, N.M.; Tishchenko, I.G.; Stanishevskii, L.S. Synthesis and some chemical transformations of 4′-aryl-4-hydroxy-4-methyl-4,5,4′,5′-tetrahydro-and 4,5-dihydro-3,3′-dipyrazolyls. *Chem. Heterocycl. Compd.,* **1984**, *20*(2), 185-188.
[http://dx.doi.org/10.1007/BF00506290]

[51]   Khalafallah, A.K. Synthesis and Reactions of 3-Formyl-1-Phenyl-5-[2-Thiazolo) Monoazamethine]Pyrazole. *Asian J. Chem.,* **1996**, *8*(4), 751.

[52]   Shawali, A.S.; Farghaly, T.A.; Aldahshoury, A.I. An efficient synthesis of functionalized 3-(hetaryl) pyrazoles. *ARKIVOC,* **2010**, *2010*(ix), 19-30.
[http://dx.doi.org/10.3998/ark.5550190.0011.903]

[53]   Shawali, A.S.; Farghaly, T.A.; Al-Dahshoury, A.R. Synthesis, reactions and antitumor activity of new β-aminovinyl 3-pyrazolyl ketones. *ARKIVOC,* **2009**, (xiv), 88-99.

[54]   Riyadh, S.M. Enaminones as building blocks for the synthesis of substituted pyrazoles with antitumor and antimicrobial activities. *Molecules,* **2011**, *16*(2), 1834-1853.
[http://dx.doi.org/10.3390/molecules16021834] [PMID: 21343888]

[55]   Dawood, K.M.; Farag, A.M.; Ragab, E.A. A facile access to polysubstituted bipyrazoles and pyrazolylpyrimidines. *J. Chin. Chem. Soc. (Taipei),* **2004**, *51*(14), 853-857.
[http://dx.doi.org/10.1002/jccs.200400128]

[56]   Dawood, K.M.; Kheder, N.A.; Ragab, E.A.; Mohamed, S.N. A facile access to some new pyrazole, 1,3,4-thiadiazole, and thiophene derivatives *via* β-ketosulfones. *Phosphorus Sulfur Silicon Relat. Elem.,* **2010**, *185*(2), 330-339.
[http://dx.doi.org/10.1080/10426500902796980]

[57]   Ioannidou, H.A.; Koutentis, P.A. The conversion of isothiazoles into pyrazoles using hydrazine. *Tetrahedron,* **2009**, *65*(34), 7023-7037.
[http://dx.doi.org/10.1016/j.tet.2009.06.041]

[58]   Rist, O.; Begtrup, M. . Synthesis of a new analogue of BINOL based on a homodimer of substituted 1-hydroxypyrazole. *J. Chem. Soc. Perkin Trans.,,* **2001**, *1*(13), 1566-1568.

[59]   Vedso, P.; Begtrup, M. Synthesis of 5-substituted 1-hydroxypyrazoles through directed lithiation of 1-(benzyloxy) pyrazole. *J. Org. Chem.,* **1995**, *60*(16), 4995-4998.
[http://dx.doi.org/10.1021/jo00121a017]

[60]   Murakami, Y.; Yamamoto, T. Synthesis of cis-bis (heteroaryl) nickel (II) complexes and reductive elimination of bis (heteroaryl) products induced by protic acid. *Inorg. Chem.,* **1997**, *36*(25), 5682-5683.
[http://dx.doi.org/10.1021/ic971111a] [PMID: 11670183]

[61]   Tang, Y.; He, C.; Imler, G.H.; Parrish, D.A.; Shreeve, J.M. A C-C bonded 5,6-fused bicyclic energetic molecule: exploring an advanced energetic compound with improved performance. *Chem. Commun. (Camb.),* **2018**, *54*(75), 10566-10569.
[http://dx.doi.org/10.1039/C8CC05987J] [PMID: 30168821]

[62]   Shkineva, T.K.; Kormanov, A.V.; Boldinova, V.N.; Vatsadze, I.A.; Dalinger, I.L. Synthesis of 4,4′-dinitro-1H,1′H-[3,3′-bipyrazole]-5,5′-diamine. *Chem. Heterocycl. Compd.,* **2018**, *54*(7), 703-709.
[http://dx.doi.org/10.1007/s10593-018-2336-5]

[63]   Dalinger, I.L.; Shkineva, T.K.; Vatsadze, I.A.; Popova, G.P.; Shevelev, S.A. N-Fluoro derivatives of nitrated pyrazole-containing fused heterocycles. *Mendeleev Commun.,* **2011**, *21*(1), 48-49.
[http://dx.doi.org/10.1016/j.mencom.2011.01.020]

[64]   Paul, H.; Kausmann, A. Über die Reaktionen von einigen konjugiert ungesättigten Kohlenwasserstoffen mit Diazomethan und Diazoäthan. *Chem. Ber.,* **1968**, *101*(11), 3700-3709.
[http://dx.doi.org/10.1002/cber.19681011108]

[65]   Yakovlev, V.; Petrov, D.V.; Dokichev, V.A.; Tomilov, Y.V. Synthesis of 3-substituted pyrazoles by oxidative dehydrogenation of 4,5-dihydro-3H-pyrazoles. *Russ. J. Org. Chem.,* **2009**, *45*(6), 950-952.
[http://dx.doi.org/10.1134/S1070428009060293]

[66]   Saha, N.; Misra, A. Design, synthesis and spectroscopic characterisation of palladium (II) and platinum (II) complexes of 5,5′-dimethyl-3,3′-bipyrazole with potential anti-tumor properties. *J. Inorg. Biochem.,* **1995**, *59*(2-3), 234.
[http://dx.doi.org/10.1016/0162-0134(95)97340-V]

CHAPTER 3

Chemistry of 3,4`-Bipyrazoles

Abstract: All the possible synthetic routes to the 3,4`-bipyrazole systems were thoroughly reported. Such synthetic platforms include: cyclocondensation and 1,3-dipolar cycloaddition reactions. Many of the reported 3,4`-bipyrazoles have potent applications in the field of pharmaceutical and material science.

Keywords: 1,3-dipolar cycloaddition, 3,4`-bipyrazoles, 3,4`-bipyrazolines, Cross-coupling, Cyclocondensation, Pyrazolylhydrazones.

1. INTRODUCTION

Various 3,4`-bipyrazoles ring skeletons were reported in the literature. They are composed of either two aromatic pyrazole units or 4-pyrazolyl attached with pyrazoline at C-3 or 4-pyrazolinyl attached to pyrazole at C-3. As a result, there will be the aromatic 3,4`-bipyrazole skeleton or partially aromatic pyrazolylpyrazoline skeleton. The two pyrazole unites are connected directly with a sigma bond between the two units. A number of tautomeric forms can be constructed, as shown in Scheme (**1**). Synthesis of such 3,4`-bipyrazole skeletons was achieved *via* several synthetic routes as outlined in Scheme (**2**). Such synthetic routes include: 1) cyclocondensation of an activated 4-pyrazole ring having chalcones or 1,3-dicarbonyl functions with hydrazines; 2) 1,3-dipolar cycloaddition of pyrazolylhydrazones with activated olefins or acetylenes; 3) 1,3-dipolar cycloaddition of nitrilimines with bis-olefines with nitrilimines or diazo-alkanes; and 4) C-C cross coupling reactions of pyrazolylboronic acids with halopyrazoles or pyrazoles themselves *via* C-H activation using palladium catalysts.

The fully aromatic 3,4`-bipyrazoles and their partially aromatic ones (pyrazolylpyrazolines) are potent inhibitory active heterocycles with significant biological potentialities. The 3,4`-bipyrazole derivatives were also considered to have anticancer [1 - 5], antimicrobial [6 - 12], anti-inflammatory [13 - 19], antioxidant [20], antitubercular [21 - 23] and antimalarial activities [24]. They

Kamal M. Dawood and Ashraf A. Abbas

were found to be effective enzyme inhibitors against carbonic anhydrase inhibitory activity [25], human Tropomyosin-related kinase A (TrkA) [26 - 30], and Janus kinase (JAK1/JAK2) [31]. 3,4`-Bipyrazole-based metal coordination complexes were reported to display remarkable pharmaceuticals applications. For example, gold(III) and iridium(II) complexes of 3,4`-bipyrazoles were useful as anticancer agents [4, 5]. In addition, the palladium(II) and platinum(II) complexes of 3,4`-bipyrazoles were found to have excellent antibacterial and antifungal activities [6, 32].

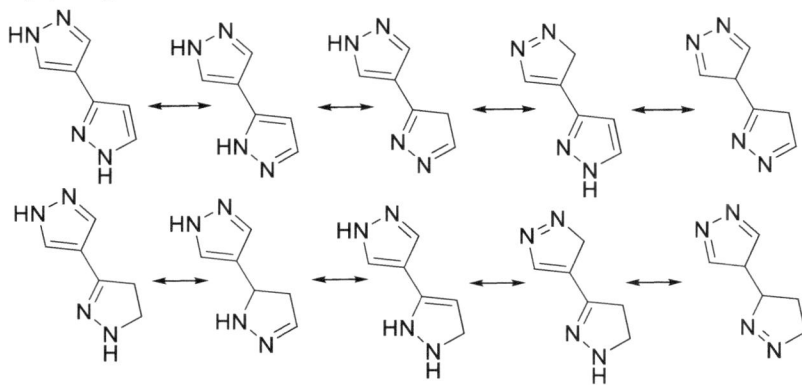

Scheme (1).   The possible direct connected 3,4`-bipyrazole derivatives

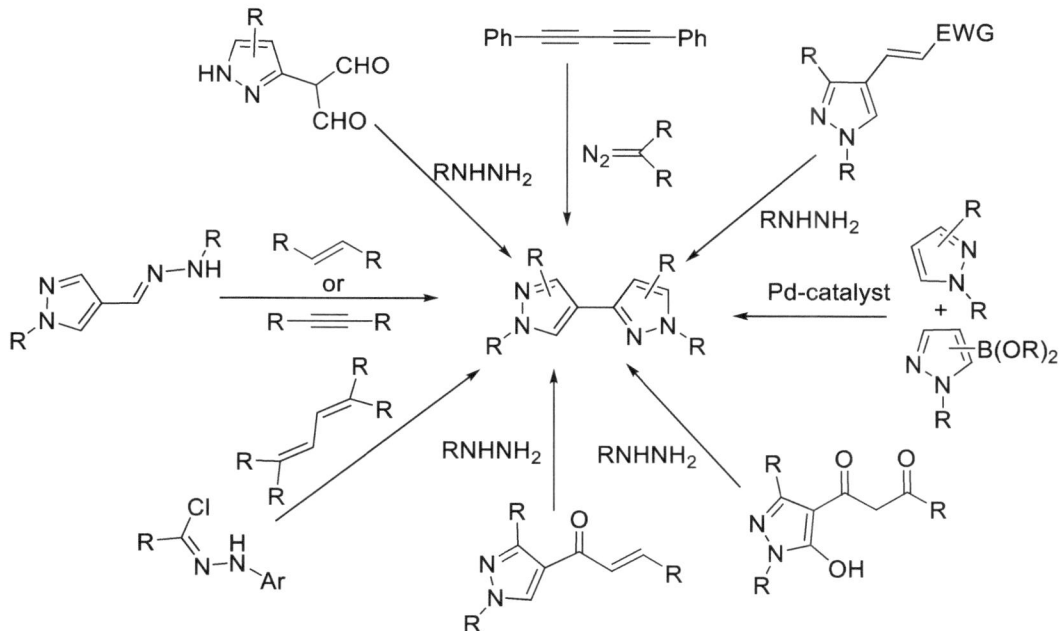

Scheme (2).   The possible synthetic routes to 3,4`-bipyrazoles systems

2. SYNTHESIS OF 3,4`-BIPYRAZOLE DERIVATIVES

2.1. From 1,3-dipolar Cycloaddition Reactions

1,3-Dipolar cycloaddition of 4-pyrazolylformylhydrazone **1** with some activated dipolarophiles such as dimethyl fumarate **3** and ethyl 3-phenylpropiolate **5** under solvent-free conditions using microwave irradiation technique resulted in the construction of the corresponding 3,4`-bipyrazoles **4** and **6**, respectively. Similar reaction of the hydrazone **1** with ethyl propiolate **7** under microwave heating at 170 °C afforded a mixture of the 3,4`-bipyrazole derivatives **8** and **9** (Scheme 3). The ^{1}H NMR analysis of structure **9** presented the following data: δ 5 1.37 (t, $J =$ 7.1 Hz, 3H, CH$_3$), 4.33 (q, $J =$ 7.1 Hz, 2H, CH$_2$), 7.35 (s, 1H, H-5'), 7.28-7.48 (m, 8H, ArH`s) 7.64 (d, $J =$ 8.6 Hz, 2H, *o*-H 1-Ph), 8.17 (s, 1H, H-3), 8.40 (s, 1H, H-5). Mechanistically, the regioselective cycloaddition process proceeded *via* the addition of the dipolarophiles **3** and **5** to the dipolar intermediate **2** followed by aromatization *via* air oxidation [33, 34]. Carrying out the 1,3-dipolar cycloaddition of the pyrazolylhydrazone **1** with dimethyl fumarate (**3**) under classical thermal heating at same temperature and reaction time on an oil bath led to the formation of the bipyrazole **4** in only 17% yield. The obtained result confirmed the advantage of microwave radiation in organic synthesis compared with classical heating.

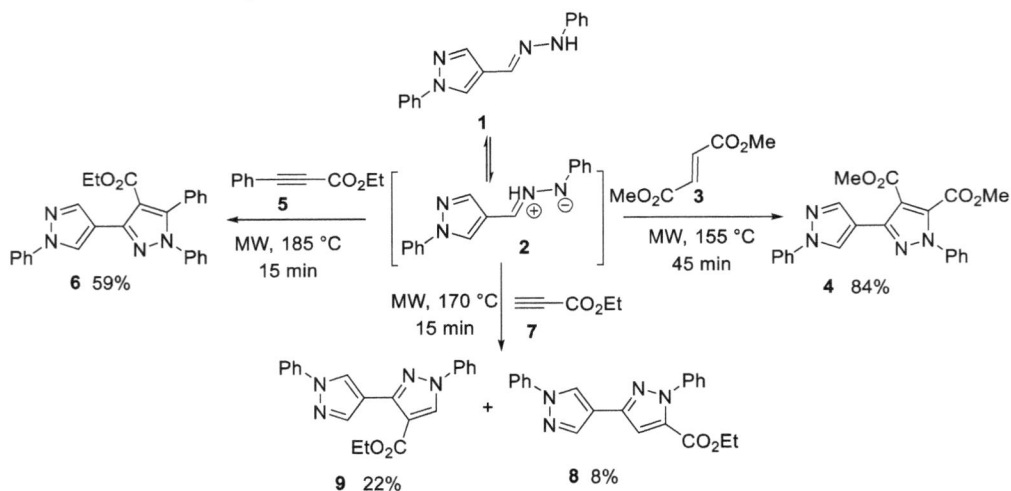

Scheme (3). Synthesis of the 3,4`-bipyrazoles **4**, **6**, **8** and **9**.

The 4-pyrazolylformylhydrazones **1** underwent similar 1,3-dipolar cycloaddition with β-nitrostyrenes **10** under solvent-free microwave irradiation condition (at 130°C for 10 min) to give a mixture of the 3,4'-bipyrazole derivatives **11** and **12** (Scheme **4**) [33, 35].

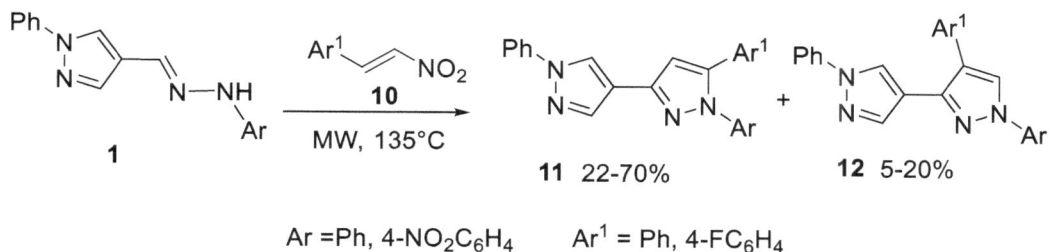

Scheme (4). Synthesis of the 3,4`-bipyrazoles **11** and **12**.

Similarly, microwave-assisted 1,3-dipolar cycloaddition reaction of the pyrazolylhydrazone **13** with ethyl 3-phenylpropiolate **5** resulted in the formation of the 3,4'-bipyrazole **15** in a high yield. The ^{1}H NMR spectrum of **15** displayed signals at δ 5 0.94 (t, J = 7.1 Hz, 3H, CH$_3$), 3.72 (s, 3H, N-CH$_3$), 4.06 (q, J = 7.1 Hz, 2H, CH$_2$), 7.27 (t, J = 7.6 Hz, 1H, p-H5'-Ph), 7.32-7.51 (m, 7H, ArH`s), 7.80 (d, J = 7.6 Hz, 2H, o-H 1-Ph), 8.30 (s, 1H, H-3), 8.79 (s, 1H, H-5). The reaction proceeded *via* the formation of the nitrilimine intermediate **14** generated by the elimination of methane from **13** (Scheme **5**). For comparison, the same cycloaddition reactions of the hydrazone **13** did not proceed under conventional thermal conditions [33, 36].

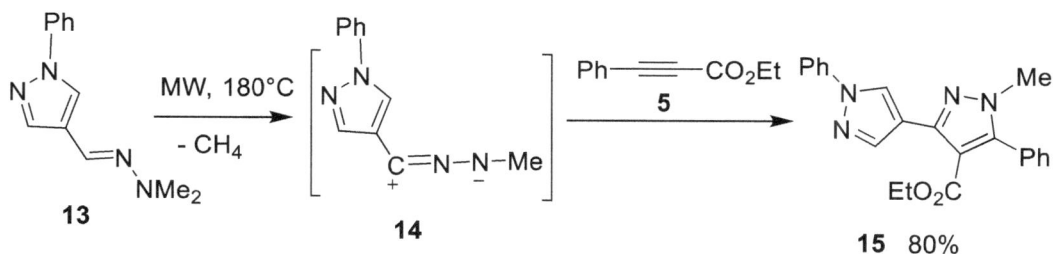

Scheme (5). Synthesis of 3,4`-bipyrazole **15**.

Treatment of the pyrazolylhydrazones **1** with *N*-bromosuccinimide (NBS) gave the pyrazolylhydrazonyl bromides **16,** which, upon treatment with Et$_3$N then the addition of C$_{60}$ under microwave irradiation conditions afforded a number of fullereno-3,4'-bipyrazole cycloadducts **18** in reasonable yields *via* the nitrilimine intermediate **17** (Scheme **6**) [37]. When the reaction was carried out under classical thermal heating, the products **18** were either not detected or obtained in very low yield (6%).

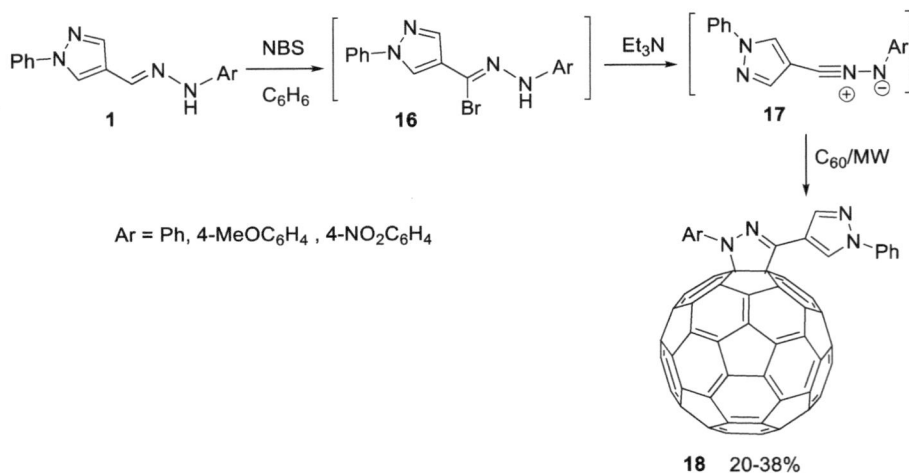

Ar = Ph, 4-MeOC$_6$H$_4$, 4-NO$_2$C$_6$H$_4$

Scheme (6). Synthesis of 3,4`-bipyrazole **18**.

The reaction of ethyl (2*Z*,4*E*)-2-cyano-5-phenyl-2,4-pentadienoate (**19**) with hydrazonoyl chloride **20**, in an equimolar ratio at ambient temperature in benzene using triethylamine as a base, afforded the regioselective cycloadduct 3,4'-bipyrazoline-5-carboxylate ester **21**. The latter bipyrazoline **21** was transformed into the pyrazolinyl-pyrazole **22** when heated at elevated temperature in DMF, *via* thermal elimination of hydrogen cyanide. In addition, concurrent decarboxylation and dehydrocyanation of the bipyrazoline **21** took place in refluxing ethanol in the presence of sodium ethoxide to give the pyrazolinyl-pyrazole **23** in a high yield. Aromatization of compounds **23** to 1,1`,3,3`,4`-pentaphenyl-3,4'-bipyrazole **24** took place upon dehydrogenation when **23** was heated in the presence of tetrachloro-1,4-benzoquinone (chloranil) (Scheme **7**) [38].

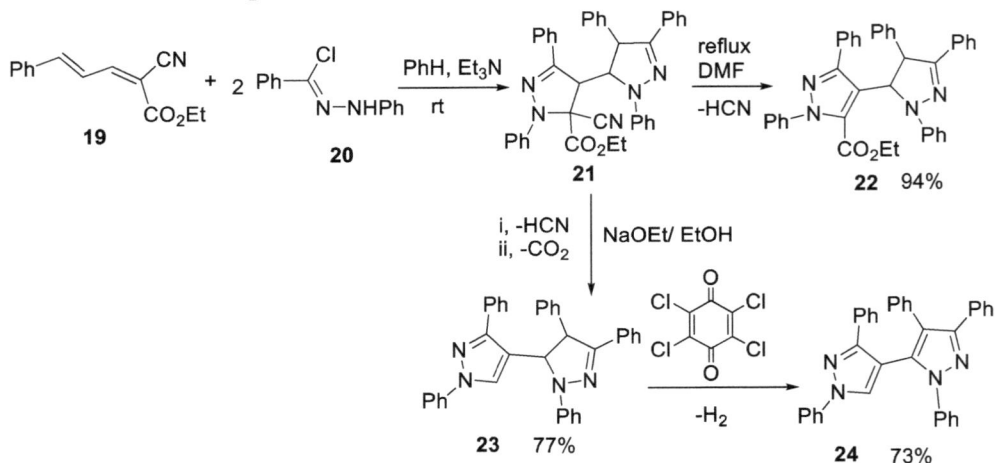

Scheme (7). Synthesis of the 3,4'-bipyrazole derivative **24**.

The reaction of diarylnitrilimines **25** (derived from the hydrazonoyl chloride **20** under the effect of Et$_3$N) with *N,N*-diethylbuta-1,3-dien-1-amine **26** (2:1 molar ratio) in refluxing benzene furnished the 3,4'-bipyrazole derivatives **29** in reasonable yields. The reaction pathway took place *via* the intermediates **27** and **28** with the elimination of diethylamine from the latter intermediate **28** (Scheme **8**) [39].

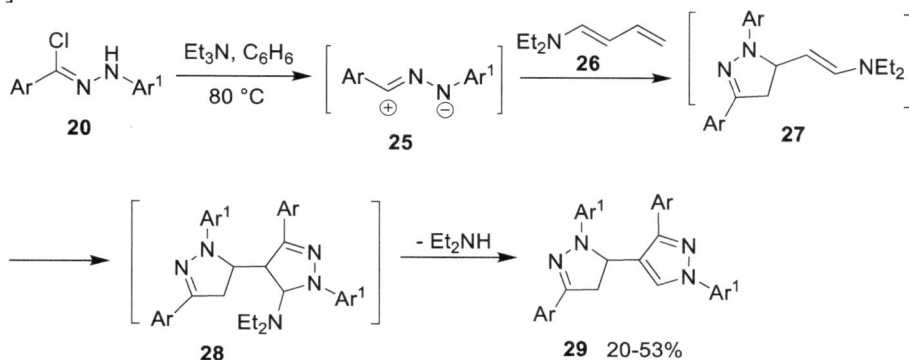

Ar/Ar^1 = Ph/Ph; Ph/4-ClC$_6$H$_4$; Ph/4-NO$_2$C$_6$H$_4$; 4-NO$_2$C$_6$H$_4$/Ph

Scheme (8). Synthesis of 3,4`-bipyrazole **29**.

Treatment of 1,2-dimethylpyrrole **31** with the hydrazonoyl chloride **30** in 1:2 molar ratio afforded a mixture of the *bis*-cycloadducts **32** and **33**. The cycloadducts **32** and **33** underwent ring transformation upon heating in ethanol solvent in the presence of hydrochloric acid to afford the 3,4'-bipyrazole **34** and 4,4'-bipyrazole **35** derivatives, respectively, in high yields (Scheme **9**) [40].

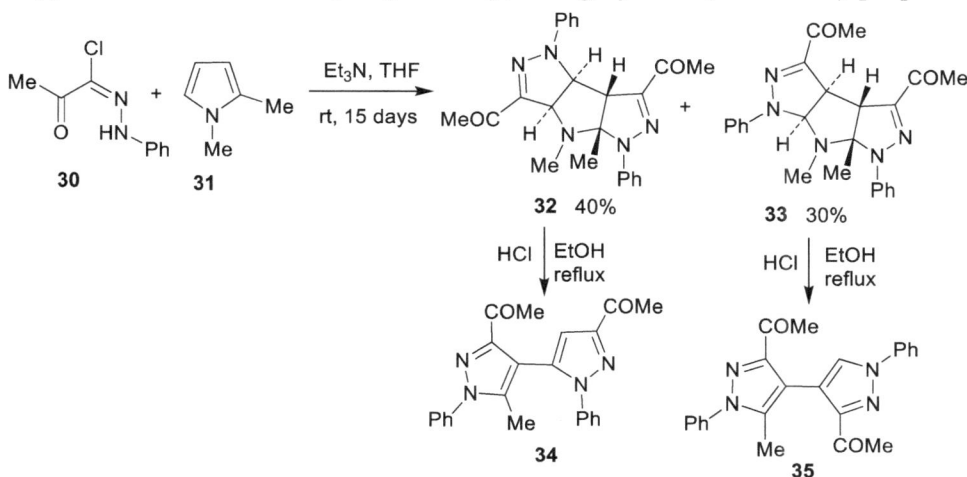

Scheme (9). Synthesis of 3,4`-bipyrazoles **34** and **35**.

2.2. From Cyclocondensation of 4-pyrazolylchalcones

The pyrazolylchalcones **38** were synthesized from the reaction of pyrazole-aldehyde **36** with an equivalent amount of acetophenones **37** in ethanolic sodium hydroxide solution at ambient temperature. Treatment of the latter chalcones **38** with hydrazine hydrate in ethanol at refluxing temperature afforded the 3,4`-bipyrazole derivatives **39** in an acceptable yield as outlined in Scheme (**10**) [13, 22].

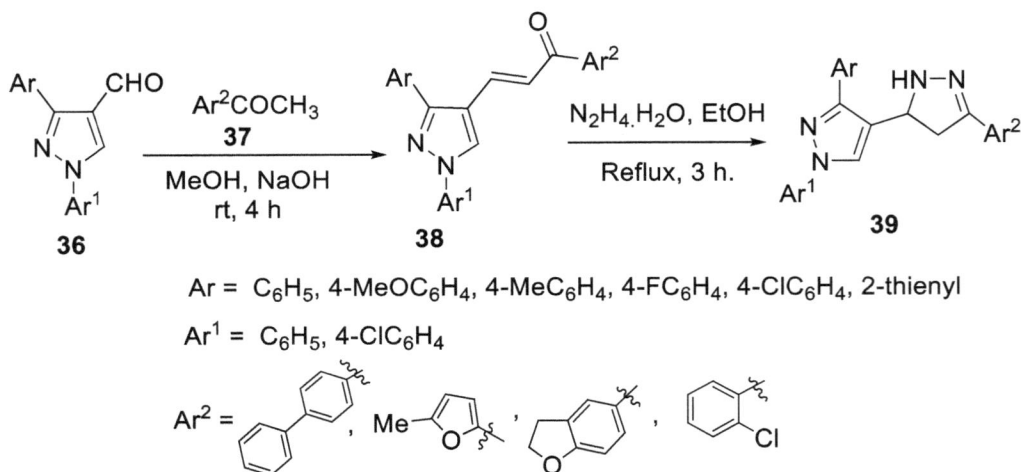

Ar = C_6H_5, 4-MeOC_6H_4, 4-MeC_6H_4, 4-FC_6H_4, 4-ClC_6H_4, 2-thienyl

Ar^1 = C_6H_5, 4-ClC_6H_4

Ar^2 =

Scheme (10). Synthesis of the 3,4'-bipyrazole derivatives **39**.

A series of pyridinoyl-bipyrazole derivatives **43** and **44** were synthesized in excellent yields using a green protocol as shown in Scheme (**11**). Thus, the reaction of the 1*H*-pyrazole-4-carbaldehydes **36a,b** with 2-acetylthiophene (**40a**) and 2-acetylfuran (**40b**) in the presence of 5% NaOH in aqueous ethanol at room temperature gave the corresponding chalcones **41a–d** in good to excellent yields. Cyclocondensation of **41a–d** with isoniazid and nicotinic acid hydrazides **42a,b** using NaOH as a catalyst at room temperature led to the formation of the corresponding bipyrazole derivatives **43a-d** in 82-88% yields and **44a–d** in 82-87% yields, respectively [41].

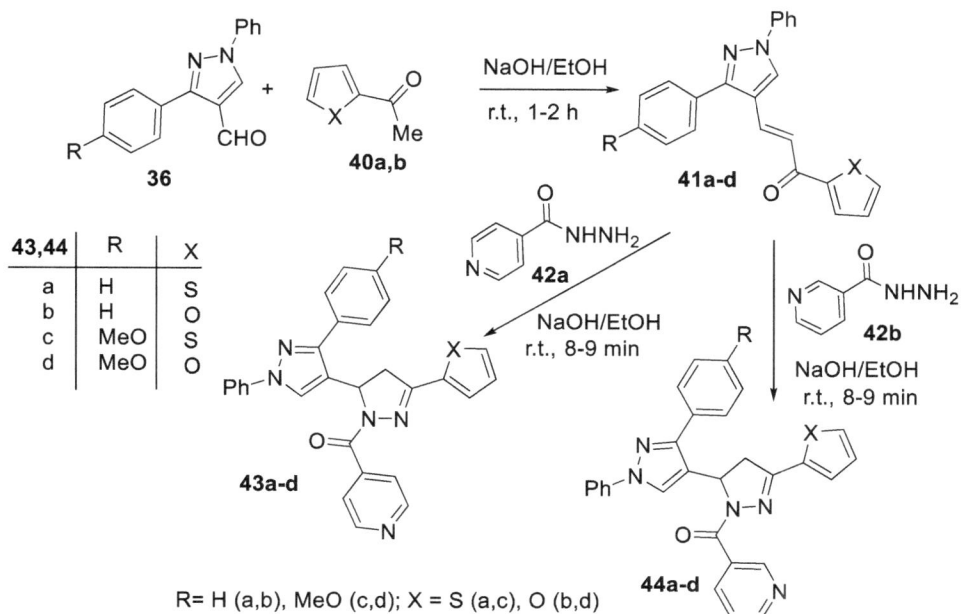

43,44	R	X
a	H	S
b	H	O
c	MeO	S
d	MeO	O

R= H (a,b), MeO (c,d); X = S (a,c), O (b,d)

Scheme (11). Synthesis of the 3,4'-bipyrazole derivatives **43** and **44**.

When pyrazole-carboxaldehyde **36** was treated with substituted acetophenone **37**, it afforded the pyrazole–chalcone conjugates **45** *via* Claisen–Schmidt condensation. The reaction of the latter chalcone **45** with semicarbazide or thiosemicarbazide in ethanol in the presence of sodium hydroxide furnished the 3,4`-bipyrazole derivatives **46** (Scheme **12**) [1, 16].

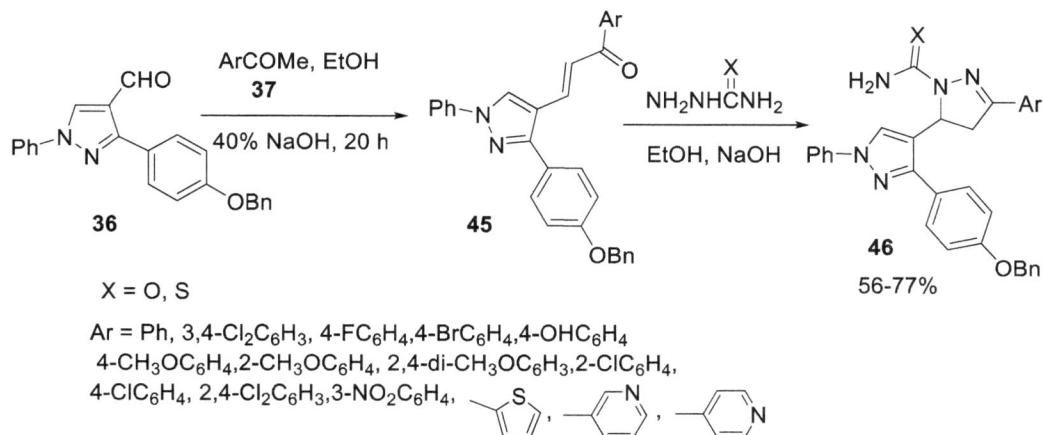

X = O, S

Ar = Ph, 3,4-Cl$_2$C$_6$H$_3$, 4-FC$_6$H$_4$,4-BrC$_6$H$_4$,4-OHC$_6$H$_4$
4-CH$_3$OC$_6$H$_4$,2-CH$_3$OC$_6$H$_4$, 2,4-di-CH$_3$OC$_6$H$_3$,2-ClC$_6$H$_4$,
4-ClC$_6$H$_4$, 2,4-Cl$_2$C$_6$H$_3$,3-NO$_2$C$_6$H$_4$,

Scheme (12). Synthesis of the 3,4'-bipyrazole derivatives **46**

Heating   3-(2-Naphthyl)-1-phenyl-1H-pyrazole-4-carbaldehyde   **36**   with acetophenones **37** gave the chalcones **45**. The latter chalcones **45** were turned into the corresponding 3,4`-bipyrazole derivatives **47** and **46** upon heating, in absolute ethanol, with hydrazines and thiosemicarbazide, respectively (Scheme 13) [20].

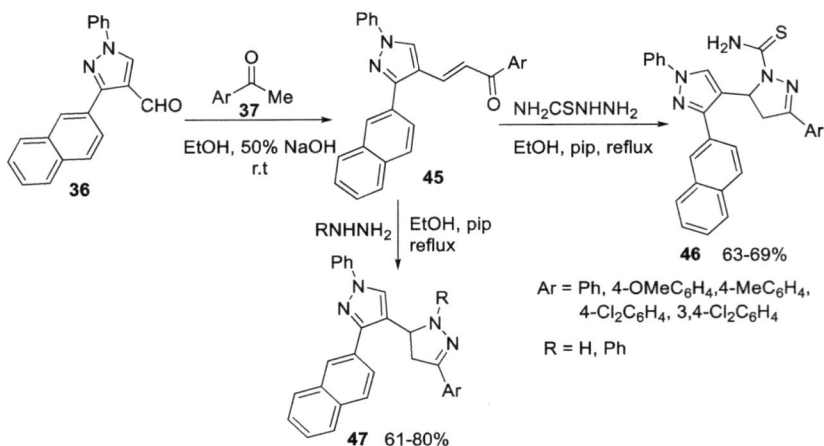

Scheme (13). Synthesis of the 3,4'-bipyrazole derivatives **46** and **47**.

The chalcones **49**, obtained from pyrazole-aldehyde **36** and disubstituted acetophenones **48**, reacted with thiosemicarbazide in EtOH/KOH at reflux condition to give the 3,4`-bipyrazole-1-thiocarboxamide derivatives **50**. The latter thiocarboxamides **50** reacted with the α-haloketones **51** under green reaction conditions in the presence of PEG-300 at room temperature to afford the 1-thiazolyl-3,4`-bipyrazole derivatives **52** in excellent yields (Scheme 14) [42].

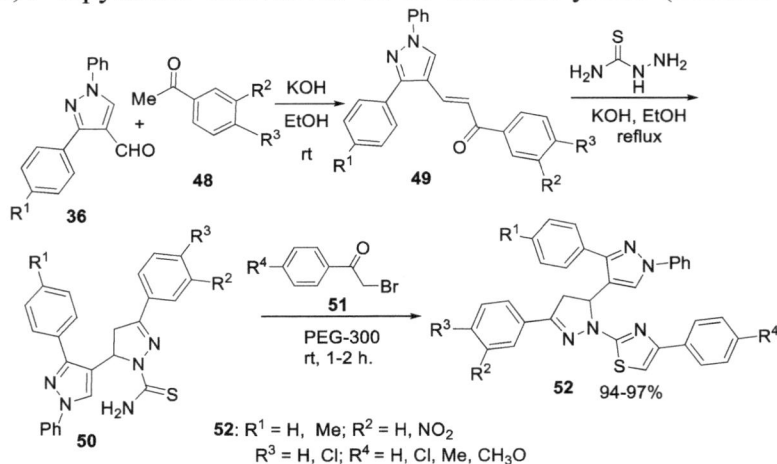

Scheme (14). Synthesis of the 3,4'-bipyrazole derivatives **50** and **52**.

The thienyl-3,4'-bipyrazole derivative **54** was reported to be obtained in a moderate yield from condensation reaction of the 3-(2-thienyl)pyrazol-4-yl) prop-2-en-1-one **53** with hydrazine hydrate in acetic acid at reflux temperature (Scheme **15**) [43].

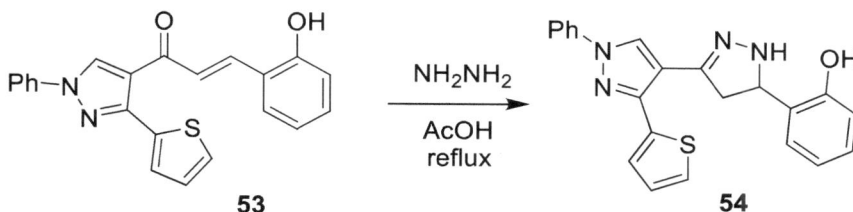

Scheme (15). Synthesis of the 3,4'-bipyrazole derivative **54**.

The reaction of 2-acetylthiophene (**40**) with an equimolar amount of pyrazole-4-carbaldehyde **55** afforded the corresponding pyrazolylchalcones **56**. The 3,4`-bipyrazole derivatives **57** were then obtained by heating the chalcones **56** with an equivalent quantity of thiosemicarbazide in ethanol at reflux in the presence of NaOH. Cyclocondensation of compounds **58** with ethyl bromoacetate in boiling ethanol gave the 3,4`-bipyrazole derivatives **58** (Scheme **16**) [10].

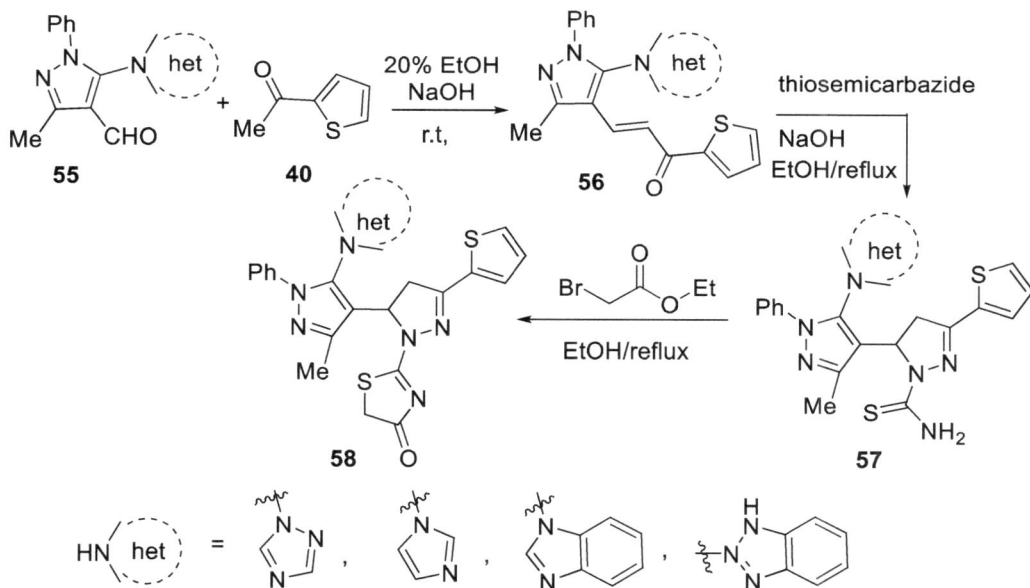

Scheme (16). Synthesis of the 3,4'-bipyrazole derivatives **58**.

Reaction of the substituted 2-acetylthiophene **40** with the pyrazole-4-carbaldehyde **59** in methanolic-KOH solution gave the chalcones **60**. The reaction of **60** with phenylhydrazine in methanolic potassium *t*-butoxide solution afforded the 3,4`-bipyrazoles **61** in good yields (Scheme **17**) [6, 32].

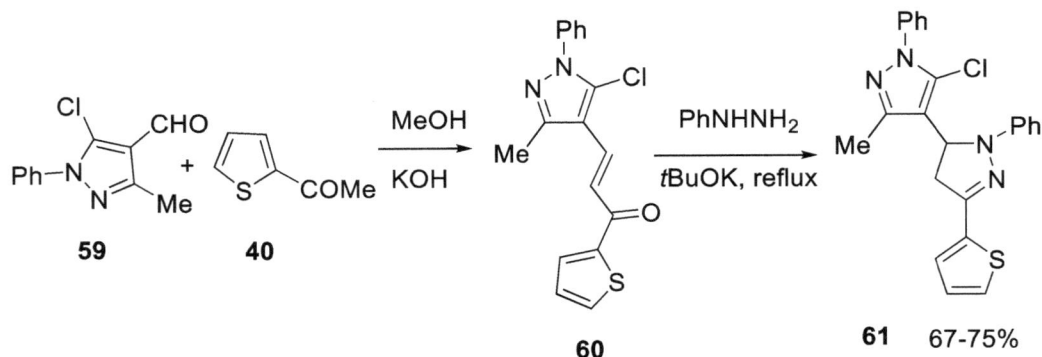

Scheme (17). Synthesis of the 3,4'-bipyrazole derivative **61**.

When the pyrazole-4-carbaldehyde derivatives **62** were treated with 4-azidoacetophenone (**63**), they afforded the corresponding chalcones **64** that underwent cyclocondensation with hydrazine in acetic acid to give the *N*-acetyl 3,4`-bipyrazole derivatives **65** in good yields (Scheme **18**) [44].

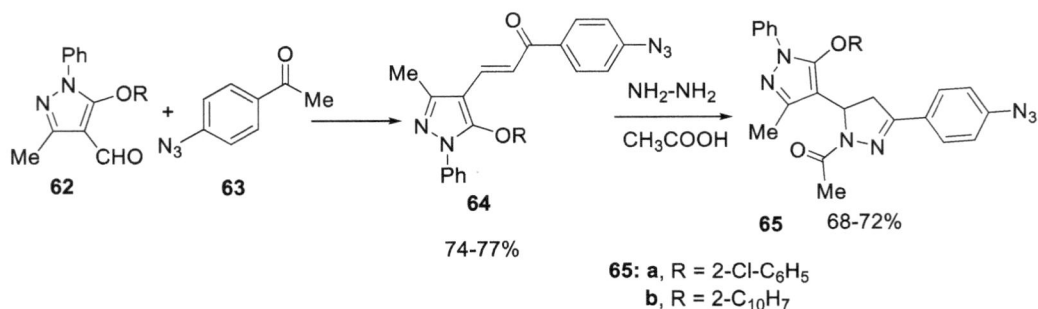

Scheme (18). Synthesis of the 3,4'-bipyrazole derivatives **65**.

Similarly, the pyrazole-4-carbaldehyde derivative **66** reacted with the acetophenone derivatives **37** in ethanol using KOH as a base to furnish the corresponding pyrazolylchalcones **67** in good yields. Cyclocondensation of the latter chalcones **67** with hydrazine gave the 3,4`-bipyrazole derivatives **68**. Heating the latter compounds **68** with formic acid or acetic anhydride led to the production of the *N*-formylated and *N*-acetylated bipyrazole derivatives **69**, respectively (Scheme **19**) [45].

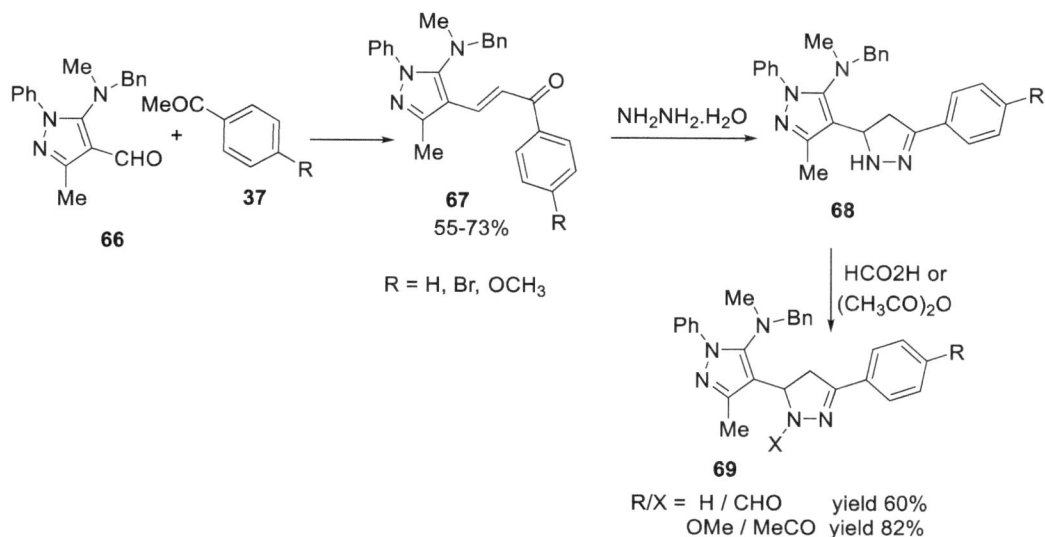

Scheme (19). Synthesis of the 3,4'-bipyrazole derivatives **68** and **69**.

The reaction of 2-acetylpyridine **71** with pyrazole-4-carbaldehyde **70** at room temperature in ethanol in the presence of NaOH gave the pyrazole-chalcones **72**. Compounds **72** were converted into the 3,4`-bipyrazole derivatives **73**, in high yields, when were treated with phenylhydrazine in ethanolic solution of potassium tertiary butoxide at reflux condition (Scheme **20**) [4].

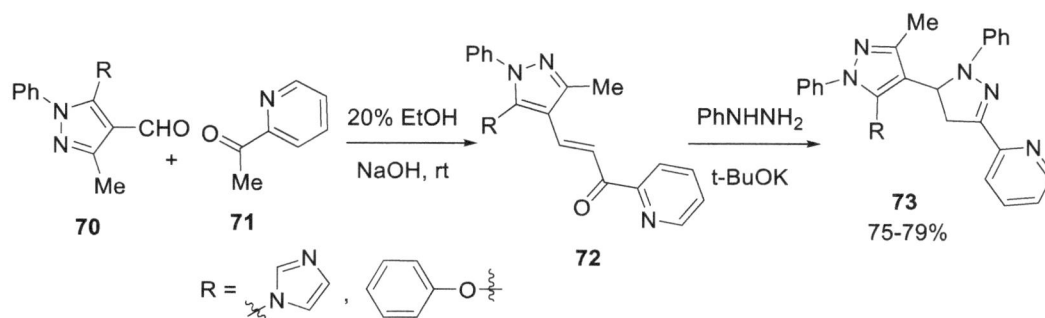

Scheme (20). Synthesis of the 3,4'-bipyrazole derivatives **73**

Heating the benzenesulfonamide-based pyrazole-4-carbaldehyde **74** with the acetophenone derivatives **37** gave the corresponding pyrazolyl-chalcones **75**. Heating the latter chalcones with 4-hydrazinobenzenesulfonamide **76** in acetic acid/ethanol mixed solvent yielded the 3,4'-bipyrazole-2,2'-dibenzenesulfonamide derivatives **77** in moderate yields (Scheme **21**) [3].

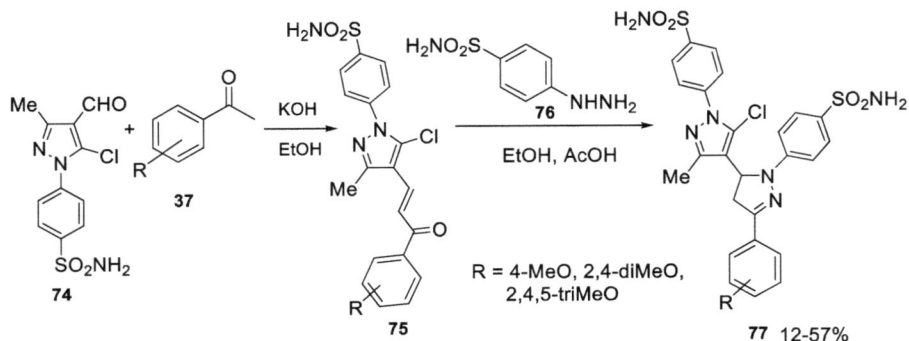

Scheme (21). Synthesis of the 3,4'-bipyrazole derivatives **77**.

When the pyrazolyl-chalcone **78** was allowed to react with 4-hydrazinobenzenesulfonamide **76** in ethanol, it afforded the dihydro-bipyrazole derivative **79**. Treatment of compound **79** with bromine water resulted in oxidation to afford the 3,4'-bipyrazole derivative **80** in 84% yield (Scheme **22**) [46].

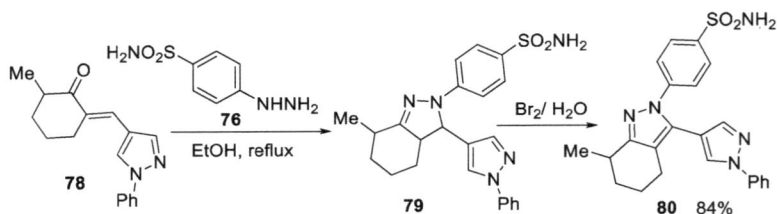

Scheme (22). Synthesis of the 3,4'-bipyrazole derivative **80**.

Heating a mixture of the pyrazolyl-chalcone derivatives **81** with the appropriate aryl hydrazines **76** led to the production of the corresponding pyrazolyl-pyrazolines **82** in high yields. Mild oxidation of the latter pyrazolinyl-pyrazoles **82** in the presence of bromine water afforded the 3,4'-bipyrazoles **83** in good yields (Scheme **23**) [15].

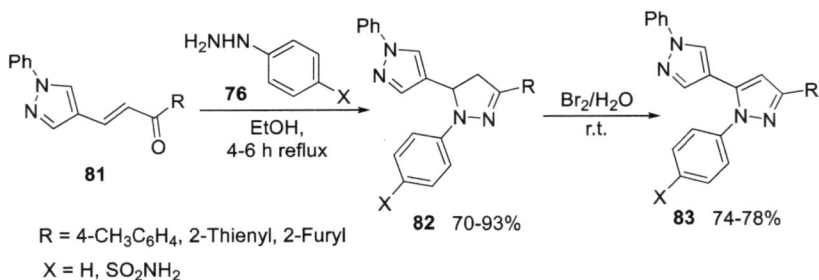

R = 4-CH$_3$C$_6$H$_4$, 2-Thienyl, 2-Furyl

X = H, SO$_2$NH$_2$

Scheme (23). Synthesis of the 3,4'-bipyrazole derivatives **83**.

In addition, heating a mixture of the (pyrazol-4-yl)chlacone derivative **85** with 2-hydrazinothiazole derivative **84** in isopropyl alcohol as solvent furnished the 3,4`-bipyrazole derivative **86** in a good yield (Scheme **24**) [47].

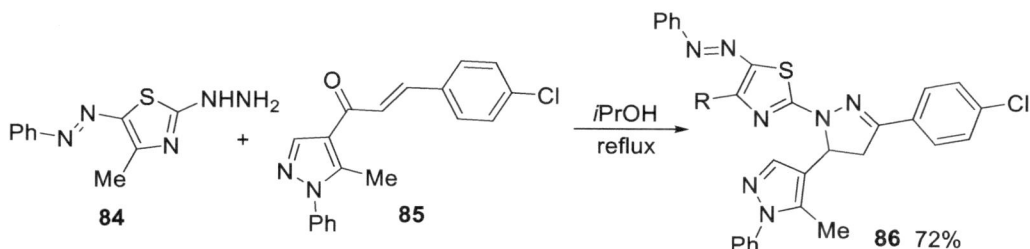

Scheme (24). Synthesis of the 3,4'-bipyrazole derivative **86**.

When 4-oxo-3-chromenecarboxaldehyde **87** was treated with 4-acetyl-5-metyl-1,2-dihydro-3-pyrazolone **88** in dry acetone and potassium hydroxide (10%), it afforded the corresponding coumarinylchalcone **89**. The latter compound **89** underwent a subsequent cyclocondensation with hydrazine hydrate under microwave irradiation conditions to give the 3,4'-bipyrazole derivative **90** in 66-82% yields (Scheme **25**) [48].

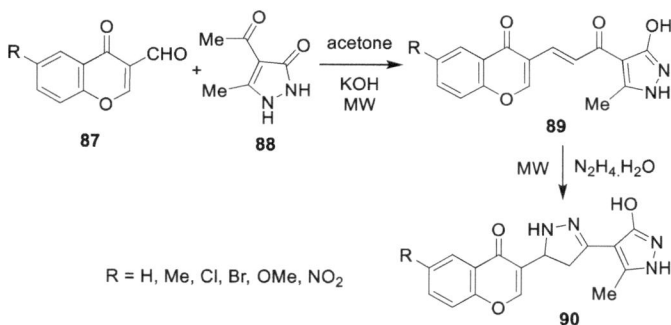

Scheme (25). Synthesis of the 3,4'-bipyrazole derivative **90**.

The reaction of 3-acetyl-4-hydroxycoumarin **91** with 3-formylchromone **87** gave the coumarinylpropenone derivative **92**. The latter was converted into the 3,4'-bipyrazole derivatives **96** when was treated with hydrazine and phenylhydrazine. Mechanistically, Scheme **26** represents ring-opening of the chromone moiety in **92** by the action of hydrazine then heterocyclization to give the pyrazole ring of **95**. The reaction of the latter with another molecule of hydrazine afforded the 3,4'-bipyrazole derivatives **96** (Scheme **26**) [49].

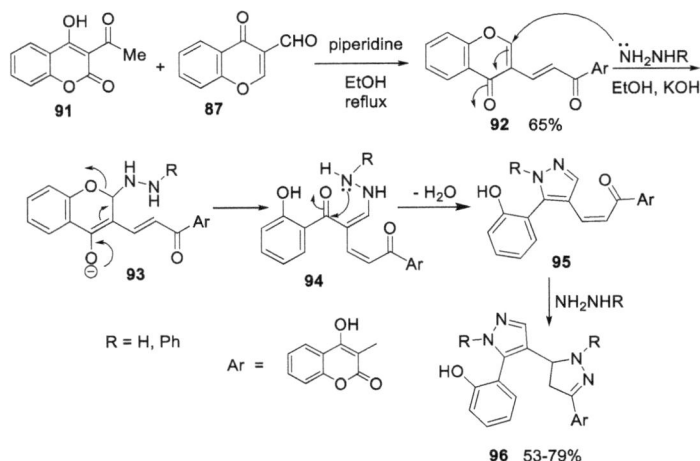

Scheme (26). Synthesis of the 3,4'-bipyrazole derivatives **96**.

A number of 1-acetyl-3,4'-bipyrazole derivatives **100** were synthesized, in good yields, from the reaction of the 3-(3-aryl-3-oxopropenyl)chromen-4-one derivatives **97** with hydrazine in acetic acid at reflux temperature. The 1-acety- -3,4'-bipyrazole derivatives **100** were oxidized by 2,3-dichloro-5,6-dicyano-p-benzoquinone (DDQ) in refluxing dioxane under nitrogen to afford the $1H,1'H$-3,4'-bipyrazole derivatives **101** in acceptable yields. Mechanistically, the formation of **100** was supposed to proceed *via* reaction of **97** with hydrazine in two possible pathways; (i) *via* 1-acetyl-3-aryl-5-(3-chromonyl)-2-pyrazoline **99** or (ii) *via* α,β-unsaturated ketone **98** intermediates. Both intermediates were treated with hydrazine to give the same products: 3,4'-bipyrazole derivatives **101** (Scheme **27**) [50, 51].

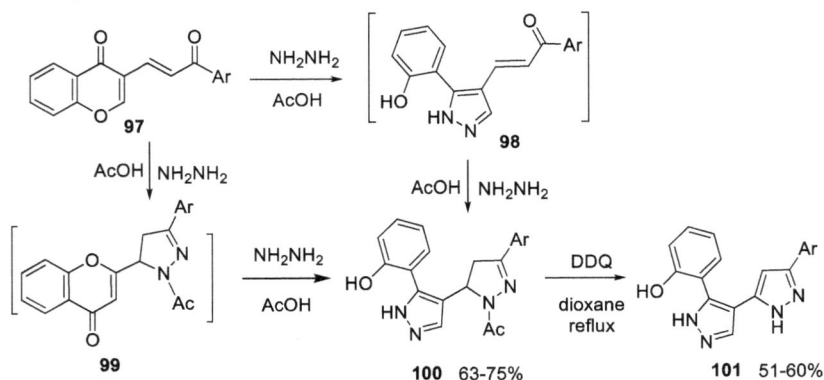

Ar = Ph, 4-MeC$_6$H$_4$, 4-MeOC$_6$H$_4$, 4-BrC$_6$H$_4$, 4-ClC$_6$H$_4$, 4-FC$_6$H$_4$, 1-naphthyl, 2-naphthyl

Scheme (27). Synthesis of the 3,4'-bipyrazole derivatives **101**.

The base catalyzed reaction of pyrazole-4-carbaldehydes **36** with 2-hydroxyacetophenones **37** afforded the corresponding propenone derivatives **102**. Treatment of the latter compounds with hydrazine and phenylhydrazine provided the 3,4'-bipyrazole derivatives **103** (Scheme 28) [52 - 54].

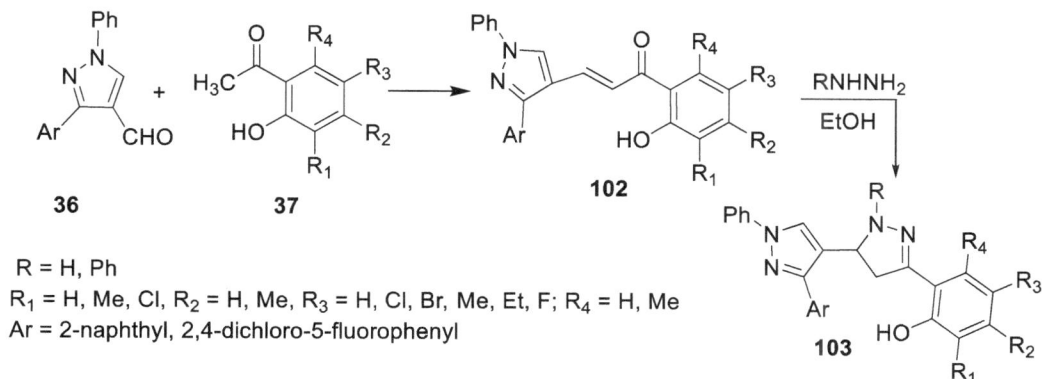

R = H, Ph
R$_1$ = H, Me, Cl, R$_2$ = H, Me, R$_3$ = H, Cl, Br, Me, Et, F; R$_4$ = H, Me
Ar = 2-naphthyl, 2,4-dichloro-5-fluorophenyl

Scheme (28). Synthesis of the 3,4'-bipyrazole derivatives **103**.

The reaction of pyrazole-4-carboxaldehyde **36** with acetophenones **37** in ethanol, at 50 °C in the presence of sodium hydroxide, generated the 4-pyrazolylpropenones **104** in high yields. The latter propenones **104** were turned into the corresponding 3,4'-bipyrazole derivatives **105** upon heating with hydrazines (Scheme **29**) [55 - 57].

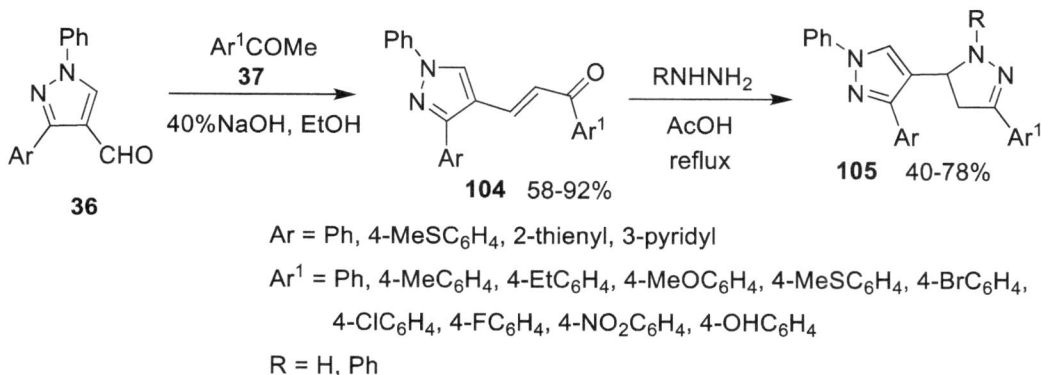

Ar = Ph, 4-MeSC$_6$H$_4$, 2-thienyl, 3-pyridyl

Ar^1 = Ph, 4-MeC$_6$H$_4$, 4-EtC$_6$H$_4$, 4-MeOC$_6$H$_4$, 4-MeSC$_6$H$_4$, 4-BrC$_6$H$_4$,

4-ClC$_6$H$_4$, 4-FC$_6$H$_4$, 4-NO$_2$C$_6$H$_4$, 4-OHC$_6$H$_4$

R = H, Ph

Scheme (29). Synthesis of the 3,4'-bipyrazole derivatives **105**.

Aldol condensation of acetophenones **37** with the 5-phenoxypyrazole-4-carboxaldehyde **106** provided the corresponding chalcone derivatives **107**. Treatment of the chalcones **107** with hydrazine in refluxing glacial acetic acid afforded the 3,4'-bipyrazole derivatives **108** in good yields (Scheme 30) [58, 59].

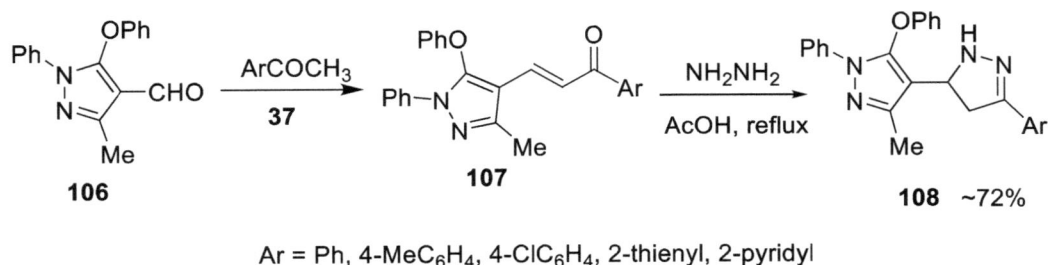

Ar = Ph, 4-MeC$_6$H$_4$, 4-ClC$_6$H$_4$, 2-thienyl, 2-pyridyl

Scheme (30). Synthesis of the 3,4'-bipyrazole derivatives **108**.

The reaction of 4-formylantipyrine **109** with acetophenone derivatives **37** led to production of the corresponding chalcones **110**. The latter chalcones were smoothly cyclized when treated with hydrazine hydrate in refluxing ethanol to give the corresponding 3,4'-bipyrazole derivatives **111** in good yields. The reaction of **111** with acetic anhydride at reflux in the presence of pyridine yielded the *N*-acylated 3,4'-bipyrazole derivatives **112** (Scheme **31**) [60].

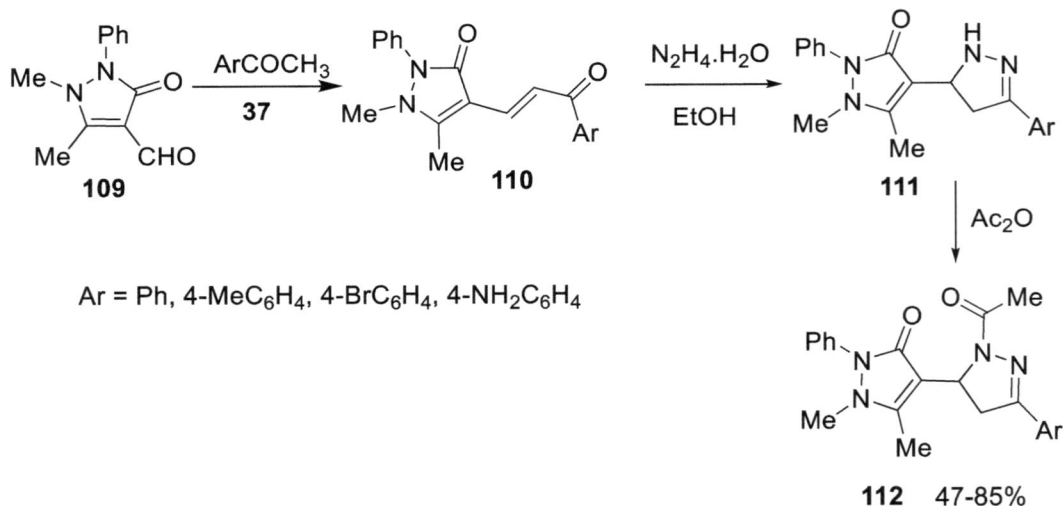

Ar = Ph, 4-MeC$_6$H$_4$, 4-BrC$_6$H$_4$, 4-NH$_2$C$_6$H$_4$

112   47-85%

Scheme (31). Synthesis of the 3,4'-bipyrazole derivatives **112**.

Mixing 1-(pyrazol-4-yl)-2-propen-1-ones **113** and arylhydrazines in DMSO solvent at reflux condition resulted in the formation of the pyrazolylpyrazoline derivatives **114** in very good yields. Treatment of the pyrazolylpyrazolines **114** with iodobenzene diacetate in dichloromethane resulted in a dehydrogenation process giving a mixture of 3,4`-bipyrazol-5`-ols **115** and their *N*-acetyl derivatives **116** in moderate yields, respectively (Scheme **32**) [9].

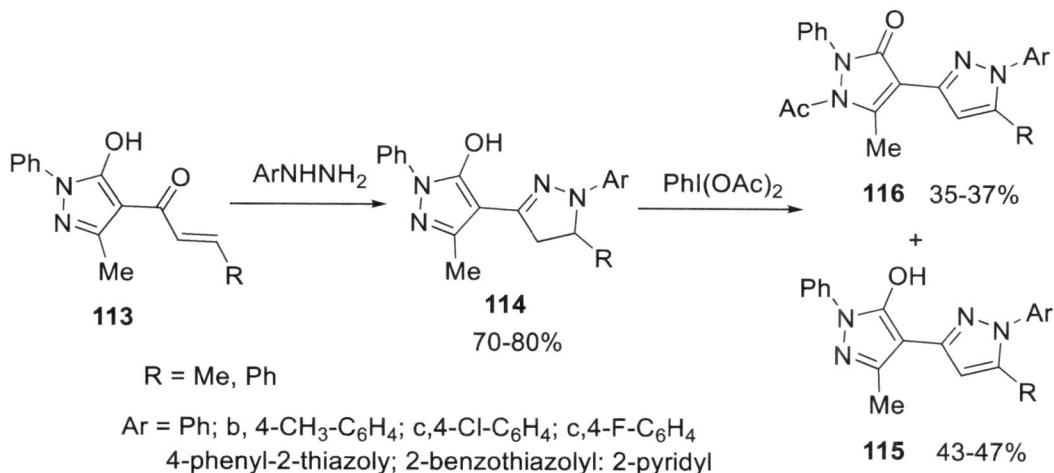

Scheme (32). Synthesis of the 3,4'-bipyrazole derivatives **115,116**.

The reaction of 4-acetylantipyrine **117** with aromatic aldehydes provided the corresponding antipyrinylchalcones **118**. The latter compounds upon heating with hydrazine hydrate afforded the 3,4'-bipyrazole derivatives **119** (Scheme **33**) [61].

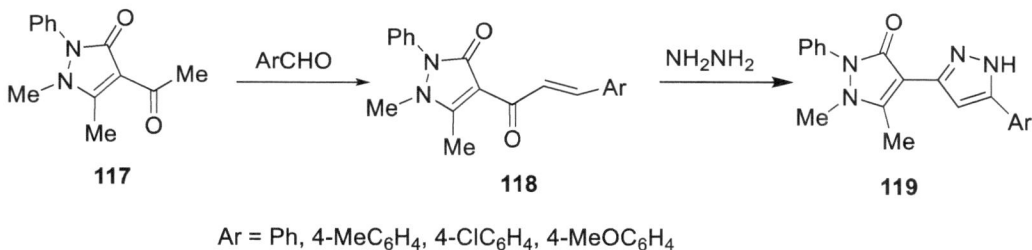

Ar = Ph, 4-MeC$_6$H$_4$, 4-ClC$_6$H$_4$, 4-MeOC$_6$H$_4$

Scheme (33). Synthesis of the 3,4'-bipyrazole derivatives **119**.

Heating a mixture of 1-isonicotinoylpyrazole-4-carboxaldehyde **120** with acetophenone derivatives **37** in ethanol under microwave irradiation in the presence of sodium hydroxide furnished the pyrazolylchalcones **121**. The reaction of **121** with isonicotinic acid hydrazide **122** under microwave irradiation condition afforded the 1,1'-diisonicotinoyl-3,4'-bipyrazole derivatives **123** in high yields (Scheme **34**) [62].

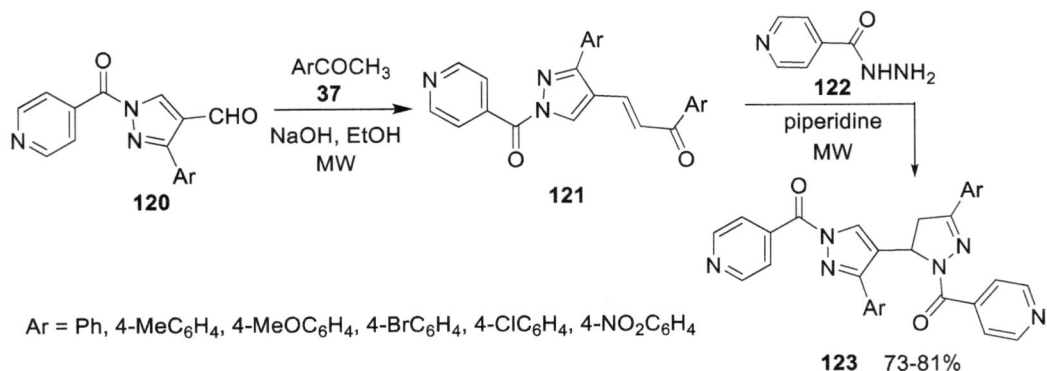

Ar = Ph, 4-MeC$_6$H$_4$, 4-MeOC$_6$H$_4$, 4-BrC$_6$H$_4$, 4-ClC$_6$H$_4$, 4-NO$_2$C$_6$H$_4$

Scheme (34). Synthesis of the 3,4'-bipyrazole derivatives **123**.

Treatment of the coumarinyl-pyrazole-4-carbaldehyde derivative **124** with the appropriate acetophenone derivatives **37** in ethanol under basic conditions yielded the chalcones **125**. The latter chalcones **125** reacted with arylhydrazines in refluxing acetic acid-ethanol mixed solvent to afford the coumarin-based 3,4`-bipyrazole derivatives **126** in moderate to excellent yields (Scheme 35) [63].

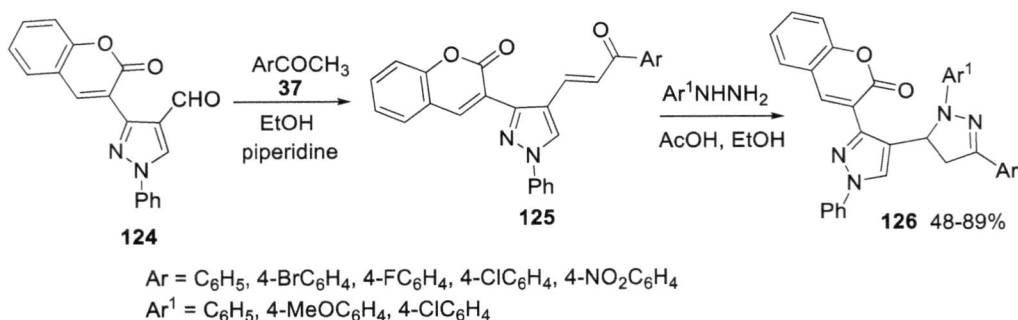

Ar = C$_6$H$_5$, 4-BrC$_6$H$_4$, 4-FC$_6$H$_4$, 4-ClC$_6$H$_4$, 4-NO$_2$C$_6$H$_4$
Ar^1 = C$_6$H$_5$, 4-MeOC$_6$H$_4$, 4-ClC$_6$H$_4$

Scheme (35). Synthesis of the 3,4'-bipyrazole derivatives **126**.

Dhinoja *et al.* reported the synthesis of a large number of 3,4`-bipyrazole derivatives **130** in high yields, as shown in Scheme (**36**). The coumarinyl chalcones **129** were prepared by the reaction of acetylcoumarin derivatives **127** with formyl pyrazoles **128**. The reaction of **129** with hydrazine hydrate gave the 3,4`-bipyrazole derivatives **130** as shown in Scheme (**36**) [8].

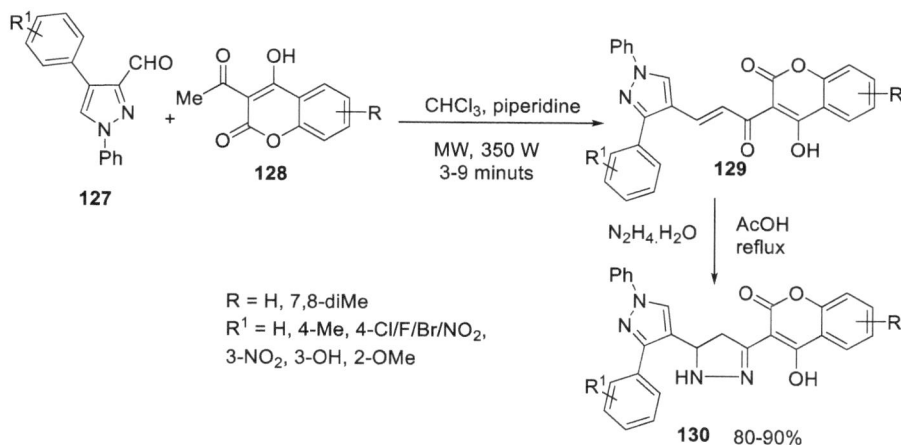

Scheme (36). Synthesis of the 3,4'-bipyrazole derivatives **130**.

Synthesis of the 3,4'-bipyrazoles **132** was carried out employing one-pot three-component procedure under three activating modes; thermal heating, ultrasound and microwave irradiation conditions, as shown in Scheme (**37**). Thus, heating an equimolar mixture of the pyrazole-4-carbaldehyde **36**, 1-(1-methoxy-2-naphthyl)ethanone **131** and hydrazines under thermal heating, ultrasound or microwave irradiation in the presence of basic alumina afforded the 3,4'-bipyrazoles **132**. Similarly, the reaction of the pyrazole-4-carbaldehyde **36** with 1-(2-methoxy-1-naphthyl)ethanone **133** and hydrazines under typical reaction conditions provided the 3,4'-bipyrazoles **134** (Scheme **37**) [64]. Heating mode had a great effect on the reaction outcomes. A comparative study between the effect of conventional heating, ultrasonic and microwave irradiations was reported as outlined in Table **1**. It was clear that the microwave irradiation technique was advantageous and superior over the other heating modes, where shorter reaction times and high productivity were observed for all reaction substrates.

Table 1. Synthesis of bipyrazoles 132a-m and 134a-k under different activation modes.

Comp.	R	Ar	Reaction Time			%Yield		
			Conv[a] (h)	US[b] (h)	MW[c] (min)	Conv	US	MW
132a	H	Ph	4.5	2.5	3.5	63	71	80
132b	H	4-Me-C_6H_5	4.5	2	3	66	74	84
132c	H	4-MeO-C_6H_5	4	2	3	65	73	85
132d	H	4-Cl-C_6H_5	5	3	3.5	62	70	83
132e	H	4-Br-C_6H_5	5	3	4	65	72	80
132f	H	4-NO_2-C_6H_5	4	3	43	60	70	82

(Table 1) cont.....

Comp.	R	Ar	Reaction Time			%Yield		
			Conv^a (h)	US^b (h)	MW^c (min)	Conv	US	MW
132g	H	thienyl	5	2.5	4	65	76	85
132h	Ph	Ph	4.5	3	3	61	72	80
132i	Ph	4-Me-C$_6$H$_5$	4.5	2.5	3	63	74	82
132j	Ph	4-MeO-C$_6$H$_5$	5	2.5	4	64	75	83
132k	Ph	4-Br-C$_6$H$_5$	5	3	4	60	72	82
132l	Ph	4-NO$_2$-C$_6$H$_5$	4.5	3	3.5	62	70	80
132m	Ph	thienyl	4.5	2.5	4	64	76	84
134a	H	Ph	4.5	3	3.5	60	72	84
134b	H	4-Me-C$_6$H$_5$	4.5	2.5	3.5	63	75	88
134c	H	4-MeO-C$_6$H$_5$	4	2.5	3.5	64	78	90
134d	H	4-Br-C$_6$H$_5$	5	3	4	62	74	86
134e	H	4-NO$_2$-C$_6$H$_5$	5	3	4	60	72	84
134f	H	thienyl	4	2.5	3	66	76	90
134g	Ph	Ph	5	3	3.5	62	70	83
134h	Ph	4-Me-C$_6$H$_5$	4.5	2.5	3.5	64	74	85
134i	Ph	4-MeO-C$_6$H$_5$	4.5	2.5	3.5	64	75	88
134j	Ph	4-Br-C$_6$H$_5$	5	3	4	61	74	86
134k	Ph	4-NO$_2$-C$_6$H$_5$	5	3	4	60	70	84

^{a}Conventional heating, ^{b}Ultrasound irradiation, ^{c}Microwave irradiation.

2.3. From Cyclocondensation of 4-pyrazolyl-bifunctional Side-arm with Hydrazines

Heating the diketopyrazoles **135** with hydrazine hydrate or aryl hydrazines in ethanol afforded the corresponding 3,4`-bipyrazole derivatives **136** and **138** in good yields. When the latter 3,4`-bipyrazole derivatives **136** and **138** were treated with phenyl isocyanate in dry acetone in the presence of potassium carbonate, they produced the corresponding thiocarbamoyl and thioureido 3,4`-bipyrazole **137** and **139** derivatives, respectively (Scheme **38**) [2].

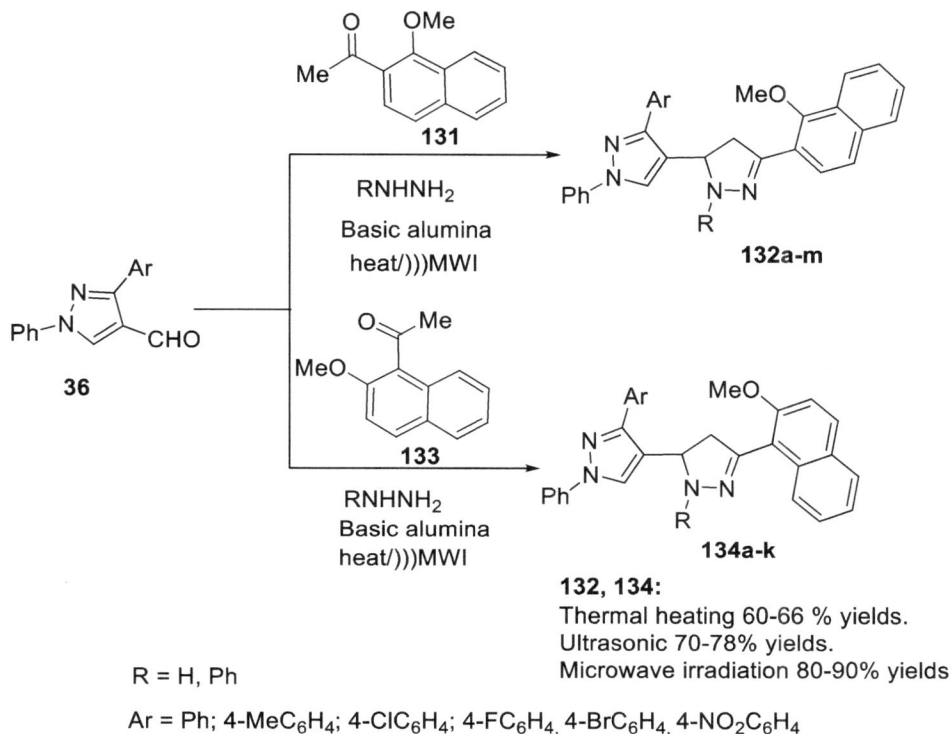

R = H, Ph

Ar = Ph; 4-MeC$_6$H$_4$; 4-ClC$_6$H$_4$; 4-FC$_6$H$_4$, 4-BrC$_6$H$_4$, 4-NO$_2$C$_6$H$_4$

Scheme (37). Synthesis of the 3,4'-bipyrazole derivatives **132** and **134**.

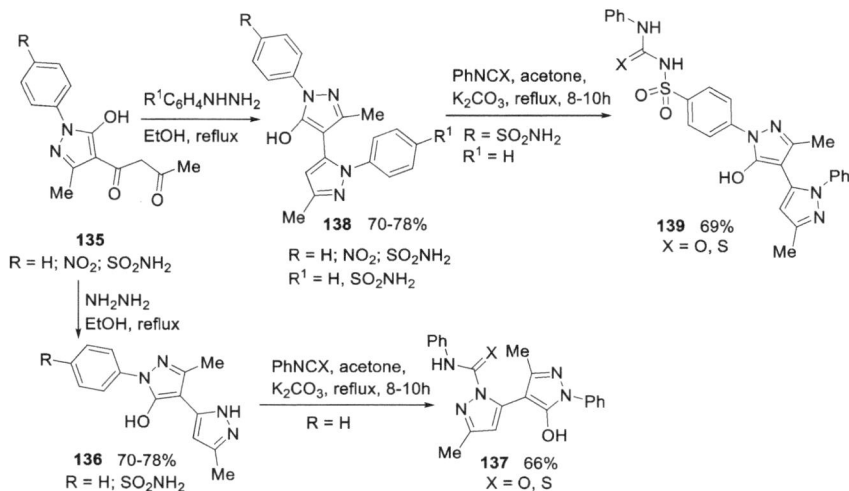

Scheme (38). Synthesis of the 3,4'-bipyrazole derivatives **136-139**.

The 1,3-diketone **143** was prepared from the reaction of dehydroacetic acid **140** with the arylhydrazine followed by heating with acetic acid as shown in Scheme (**39**). Cyclocondensation of the 1,3-diketone **143** with phenylhydrazine afforded the 3,4`-bipyrazole derivative **146** through the intermediates **141~145** according to the possible reaction mechanism illustrated in Scheme (**39**) [14].

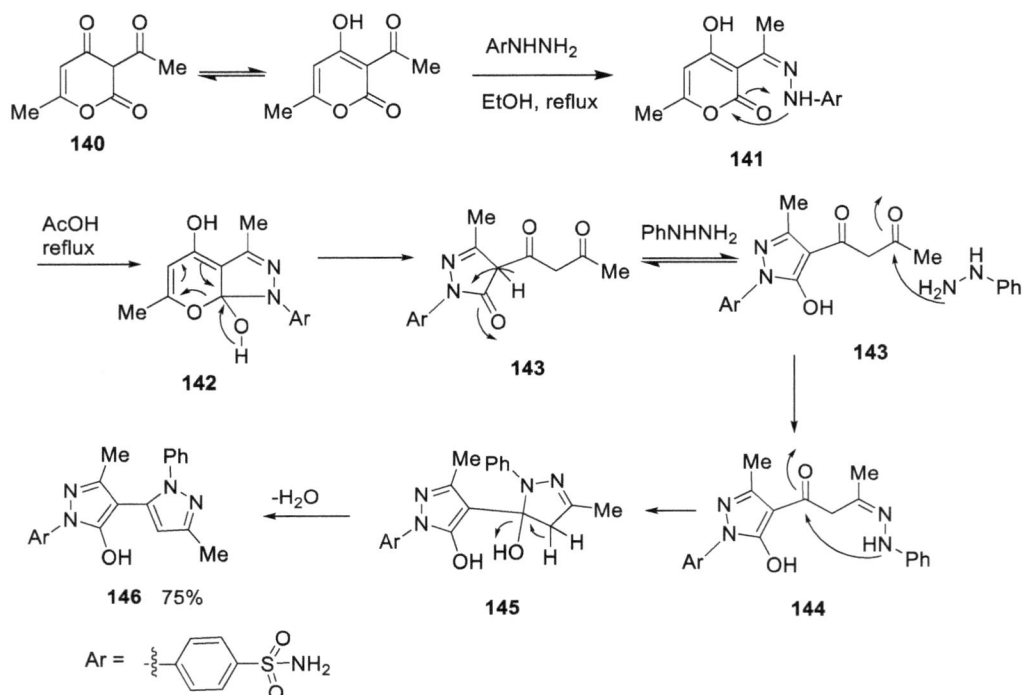

Scheme (39). Synthesis of the 3,4'-bipyrazole derivative **146**

Chemo- and regio-selective preparation of the 3,4`-bipyrazole esters **150** was performed employing 4-pyrone derivatives **147** with hydrazine hydrochloride or with methyl hydrazine acetate in refluxing ethanol according to the suggested mechanism depicted in Scheme (**40**) [65].

Next, the γ-pyrone derivative **151** reacted with two equivalents of hydrazine hydrate to give the 3,4'-bipyrazole derivative **155** as outlined in Scheme (**41**), where hydrazinloysis of ester group firstly took place then the nucleophilic attack of another hydrazine molecule *via* ring-opening and ring-closing pathway with loss of water molecule to provide the 3,4'-bipyrazole derivative **155** (Scheme **41**) [66].

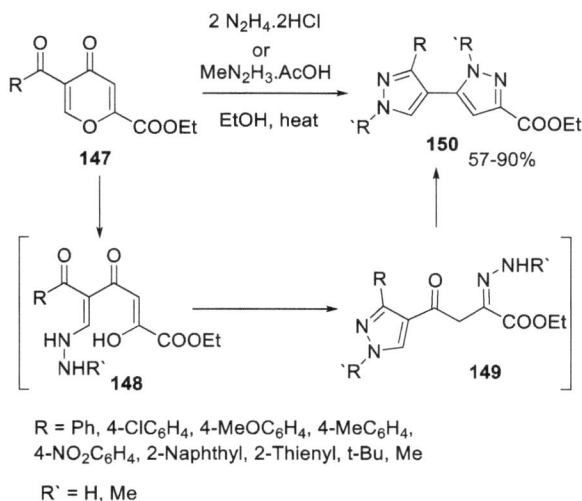

R = Ph, 4-ClC$_6$H$_4$, 4-MeOC$_6$H$_4$, 4-MeC$_6$H$_4$,
4-NO$_2$C$_6$H$_4$, 2-Naphthyl, 2-Thienyl, t-Bu, Me

R` = H, Me

Scheme (40). Synthesis of the 3,4'-bipyrazole derivatives **150**.

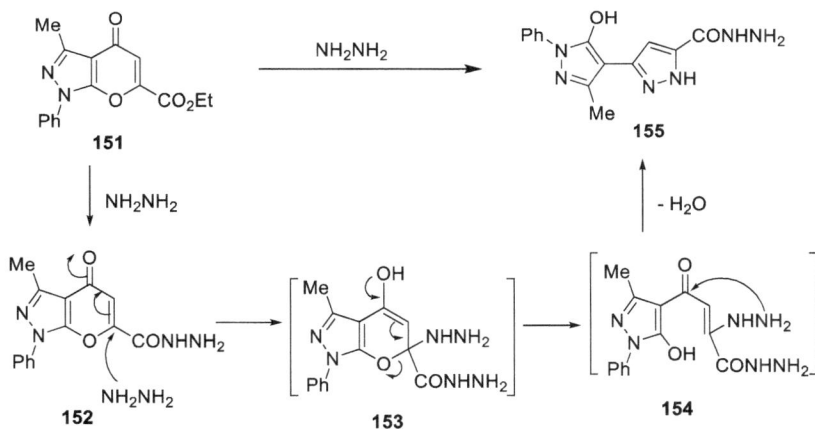

Scheme (41). Synthesis of the 3,4'-bipyrazole derivative **155**.

Heating the 1-(pyrazol-4-yl)butane-1,3-dione **156** with hydrazine hydrate in ethanol furnished the 3,4'-bipyrazolyl-5'-ol derivative **157,** which was useful for treating the central nervous system diseases (Scheme **42**) [67].

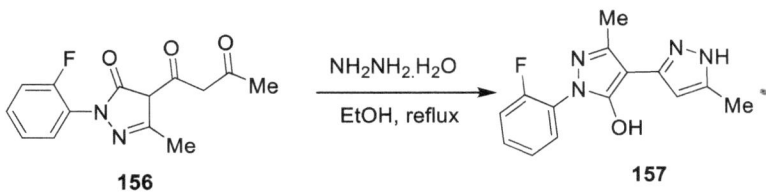

Scheme (42). Synthesis of the 3,4'-bipyrazole derivative **157**.

Reaction of dehydroacetic acid **140** (or its tautomer(with a number of hydrazine derivatives provided the corresponding hydrazones **141**, which in turn underwent rearrangement in refluxing acetic acid to form the corresponding 4-acetoacetylpyrazole derivatives **143** (Scheme **43**). The reaction of the 4-acetoacetylpyrazole derivatives **143** with aryl- or heteroaryl-hydrazines in refluxing ethanol in the presence of HCl as an acid catalyst afforded the 3,4'-bipyrazoles **158** in both their *NH*- and *OH*-tautomeric forms (**158** and **158A**) (Scheme **43**) [68 - 78].

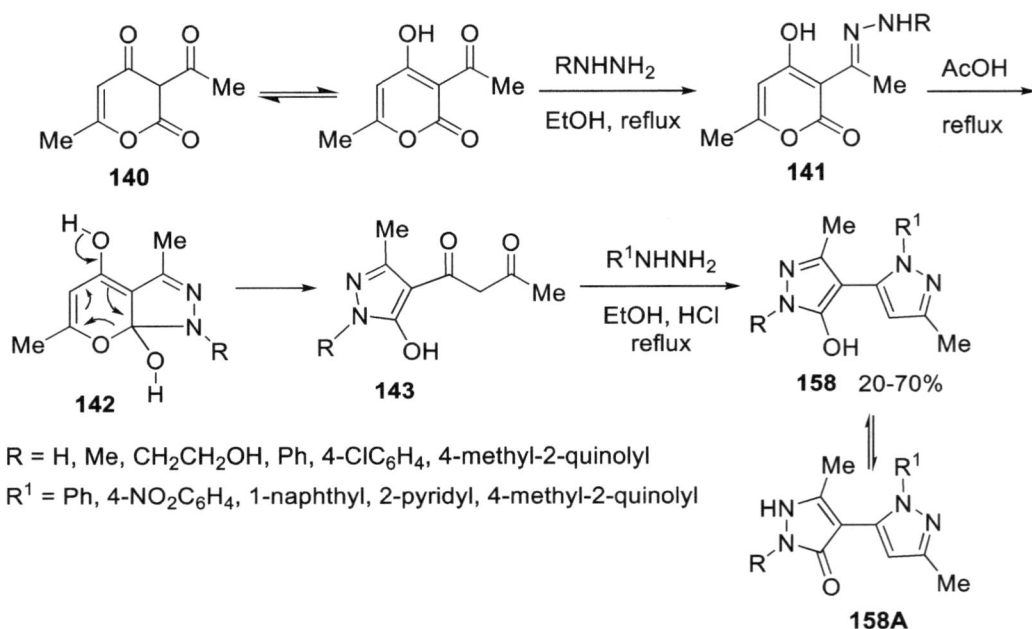

R = H, Me, CH₂CH₂OH, Ph, 4-ClC₆H₄, 4-methyl-2-quinolyl
R¹ = Ph, 4-NO₂C₆H₄, 1-naphthyl, 2-pyridyl, 4-methyl-2-quinolyl

Scheme (43). Synthesis of the 3,4'-bipyrazole derivatives **157**.

A number of the isomeric 3,4`-bipyrazoles structures (**159** and **161**) having 4-(aminosulfonyl)phenyl moiety were synthesized in good yields from the reaction of either (pyrazol-4-yl)-1,3-butanediones **143** or **160** with the appropriate arylhydrazines in EtOH/HCl under reflux condition (Scheme **44**) [79].

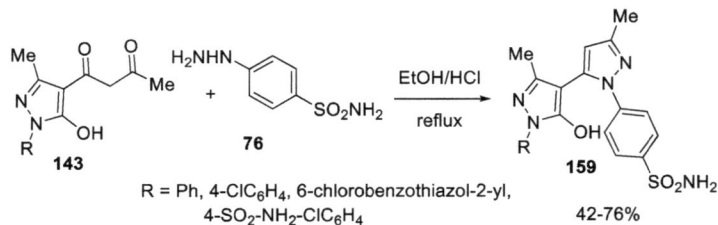

R = Ph, 4-ClC₆H₄, 6-chlorobenzothiazol-2-yl,
4-SO₂-NH₂-ClC₆H₄

(Scheme 44) cont.....

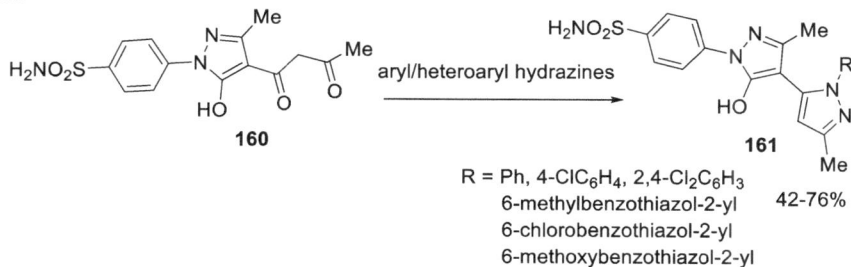

Scheme (44). Synthesis of the 3,4'-bipyrazole derivatives **159** and **161**.

Dehydroacetic acid (**140**) was also utilized in the synthesis of the 3,4'-bipyrazol-5-ol derivatives **165** as shown in Scheme (**45**). Therefore, heating dehydroacetic acid **140** with thiosemicarbazide (1:1 molar ratio) in refluxing ethanol followed by addition of the phenacyl bromide derivatives **51** afforded the arylthiazolyl-hydrazonopyran-2-one derivatives **162**. Rearrangement of the latter compounds **162** took place in refluxing ethanol/acetic acid mixed solvent to afford the pyrazol-4-ylbutane-1,3-diones **163**. The latter diones **163** were converted into the 3,4'-bipyrazol-5-ols **165** in good to excellent yields when heated with 2-hydrazinobenzothiazole **164** under the same reaction condition (Scheme **45**) [80].

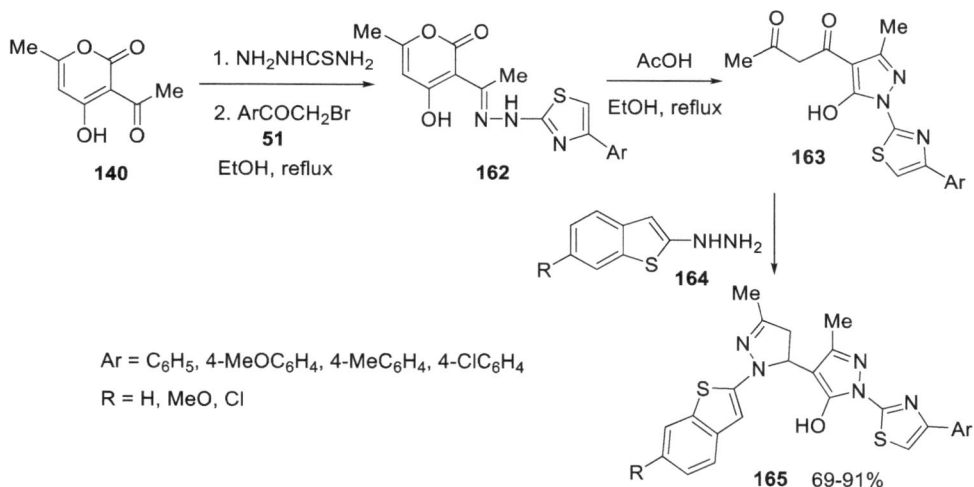

Scheme (45). Synthesis of the 3,4'-bipyrazole derivatives **165**.

When 3,5-diacetyl-2,6-dimethyl-4-pyrone **166** was allowed to react with hydrazine and phenylhydrazine at ambient temperature then acidification with sulphuric acid, it afforded the 4-acetoacetylpyrazole derivatives **169** *via* the intermediates **167** and **168** *via* loss of water then acetic acid molecules. Treatment of the products **169** with another mole of hydrazines produced the non-symmetric 3,4'-bipyrazole derivatives **170** in acceptable yields (Scheme **46**) [81].

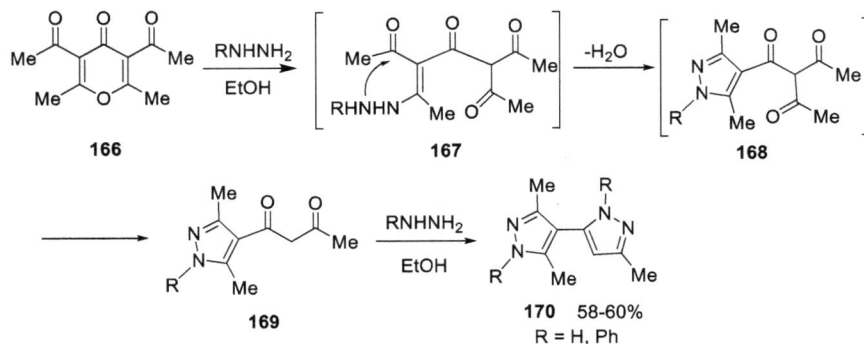

Scheme (46). Synthesis of the 3,4'-bipyrazole derivatives **170**.

Synthesis of the thiazole-based 3,4`-bipyrazole derivatives **174** was carried out through a one-pot three-component route. Therefore, when α-bromoketones **51**, thiosemicarbazide, and the diketo-pyrazole derivatives **171** were mixed together in 1:1:1 molar ratio in ethanol and heated at reflux, the corresponding 3,4`-bipyrazole derivatives **174** were obtained in high yields. The reaction proceeded *via* the possible reaction mechanism outlined in Scheme (**47**) *via* the elimination of two water molecules from the intermediates **172** and **173** [82].

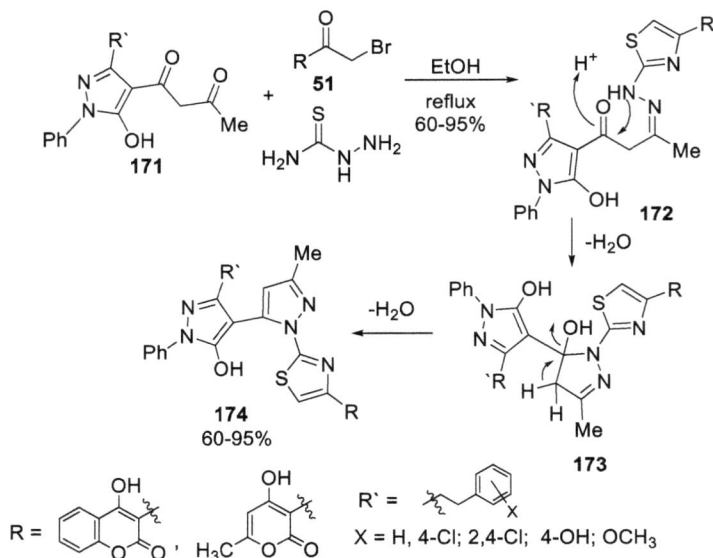

Scheme (47). Synthesis of the 3,4'-bipyrazole derivatives **174**.

The reaction of 5-pyrazolone **175** with triethyl-*ortho*-acetate [MeC(OEt)$_3$] in acetic acid at reflux gave the 4-acetylpyrazole derivative **176**. Treatment of the latter compound **176** with boron trifluoride etherate and tributyl borate then the addition of dimethylformamide-dimethylacetal (DMF-DMA) in tetrahydrofuran at ambient temperature furnished the intermediate **177**. Treatment of **177** with an equivalent amount of the appropriate arylhydrazine derivative in refluxing ethanol led to the formation of the 3,4'-bipyrazole derivatives **178a-e** in good yields (Scheme **48**) [83].

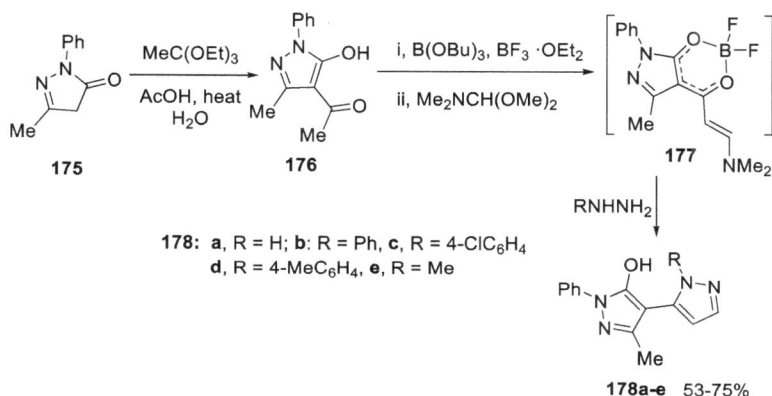

Scheme (48). Synthesis of the 3,4'-bipyrazole derivatives **178a-e**.

In a similar fashion, the thiazolone-based 3,4'-bipyrazole derivatives **179** were synthesized employing a one-pot four-component protocol of equimolar amounts of phenacyl bromides **51**, pyrazole-aldehydes **36**, thiosemicarbazide and chloroacetic acid. The reaction was accomplished under the green condition in PEG-400 (polyethylene-glycol) in the presence of bleaching earth clay (BEC) (pH 12.5, 10 wt %) as a basic catalyst at 70- 80 °C (Scheme **49**) [21].

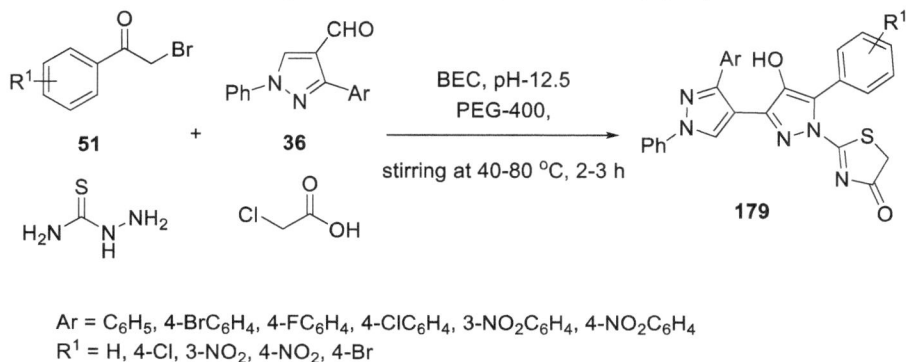

Ar = C$_6$H$_5$, 4-BrC$_6$H$_4$, 4-FC$_6$H$_4$, 4-ClC$_6$H$_4$, 3-NO$_2$C$_6$H$_4$, 4-NO$_2$C$_6$H$_4$
R^1 = H, 4-Cl, 3-NO$_2$, 4-NO$_2$, 4-Br

Scheme (49). Synthesis of the 3,4'-bipyrazole derivatives **179**.

Heating the furo[3,2-g]chromen-6-yl)prop-2-enal derivative **180** with an equimolar amount of hydrazine hydrate or phenylhydrazine in ethanol at reflux temperature afforded the 3,4`-bipyrazole derivatives **183**. The suggested mechanism is depicted in Scheme (**50**), where condensation of hydrazine with the formyl group took place concurrently with cyclization *via* elimination of HCl molecule to afford the intermediate **181**. The latter intermediate upon attack of another hydrazine molecule gave the bipyrazole **183** *via* ring opening of the γ-pyrone followed by repeated cyclization with the elimination of water molecule from the intermediate **182** (Scheme **50**) [84]. The ^{1}H NMR spectrum of the derivative **183** (R = H, R_1 = Ph) exhibited the following characteristic data: δ 3.82 s (3H, OCH$_3$), 3.95 s (3H, OCH$_3$), 6.69 (d, 1H, H$^3_{pyrazole}$, J = 6.6 Hz), 7.04 (d, 1H, H$^3_{furan}$, J = 1.8 Hz), 7.26–7.48 (m, 5H, Ph–H),7.61 d (1H, H^{4}pyrazole, J = 6.6 Hz), 7.91 (d, 1H, H$^2_{furan}$, J 1.8 Hz), 8.30 (s, 1H, H$^3_{pyrazole}$), 9.28 br.s (1H, NH exchangeable with D$_2$O), 10.46 (br.s, 1H, OH exchangeable with D$_2$O). The ^{13}C NMR of the same derivative **183** (R = H, R_1 = Ph) showed the following signals: δ 58.0 (OCH$_3$), 59.3 (OCH$_3$), 103.4, 105.8, 107.9, 112.0, 114.7, 121.5, 124.7, 126.6, 127.3, 130.2, 134.8, 138.7, 141.4, 146.5, 147.6, 149.0, 150.8, 152.7 ppm.

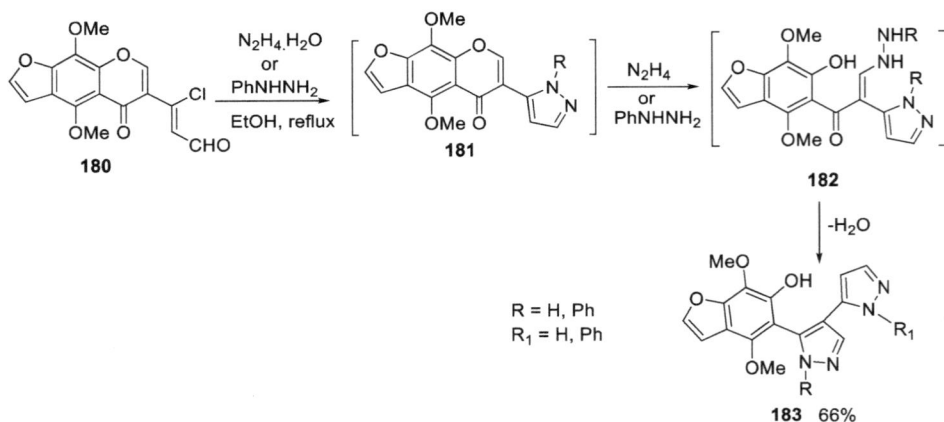

Scheme (50). Synthesis of the 3,4'-bipyrazole derivatives **183**.

Heating the 3-(pyrazol-4-yl)-3-chloropropenal **184** with arylhydrazines afforded the 3-chloro-3,4'-bipyrazole derivatives **185** (Scheme **51**) [85].

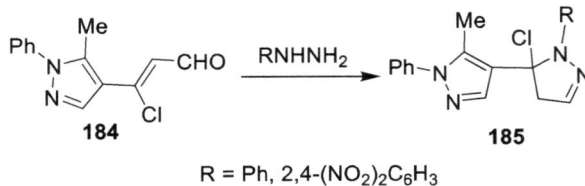

R = Ph, 2,4-(NO$_2$)$_2$C$_6$H$_3$

Scheme (51). Synthesis of the 3,4'-bipyrazole derivatives **185**.

When (Z)-3-bromo-4-(1,3-diaryl-1*H*-pyrazol-4-yl)-3-buten-2-ones **186** was allowed to react with phenylhydrazine in acetic acid at reflux temperature, it afforded the 3,4`-bipyrazole derivatives **187** in high yields (Scheme **52**) [11].

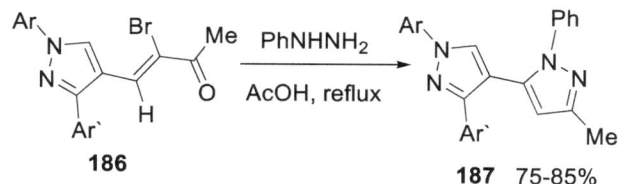

Scheme (52). Synthesis of the 3,4'-bipyrazole derivatives **187**.

Treatment of methyl cyano-(3-cyano-4,5-dihydro-2(3*H*)-furanylidene)acetate (**188**) with double molar amount of hydrazine in 1,2-dichlorobenzene furnished the 3,4'-bipyrazole derivative **190** in a high yield, *via* the isolable hydrazide intermediate **189**. Heating the bipyrazole **190** with *N,N*-dimethylformamid--dimethylacetal (DMF-DMA) led to the generation of the three-ring fused heterocycles **192** in a high yield *via* extrusion of dimethylamine from the bipyrazole intermediate **191** (Scheme **53**) [86].

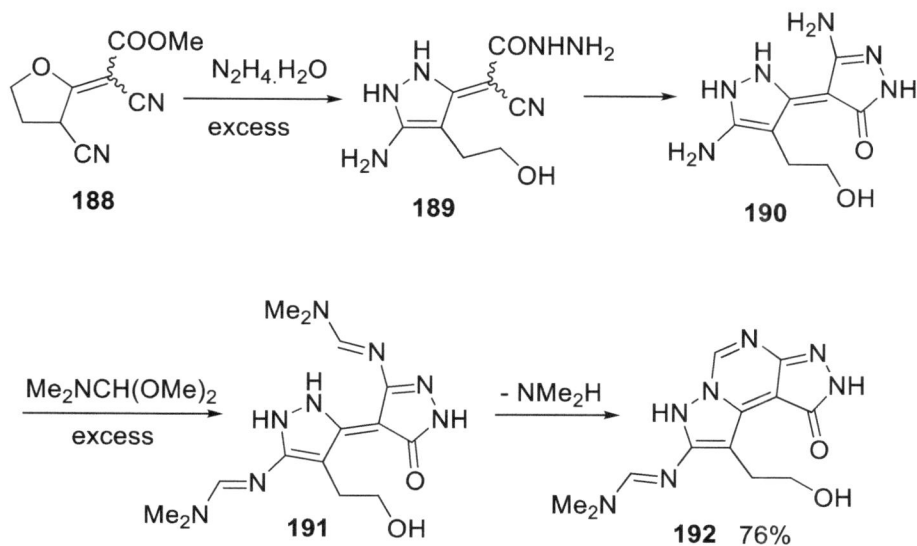

Scheme (53). Synthesis of the 3,4'-bipyrazole derivatives **190-192**.

Next, when a mixture of the (pyrazol-4-yl)methylene)malononitrile derivative **193** and the pyridazinylacetohydrazide derivative **194** was heated in dioxane in the presence of piperidine as a base, it afforded the 3,4'-bipyrazole derivative **195** in a moderate yield (Scheme **54**) [87].

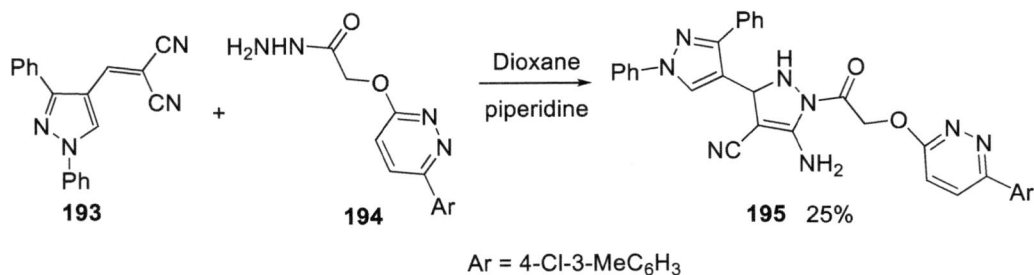

Ar = 4-Cl-3-MeC$_6$H$_3$

Scheme (54). Synthesis of the 3,4'-bipyrazole derivative **195**.

Heating 4-acetyl-1-phenylpyrazole **196** with DMF-DMA in xylene led to the production of the enaminone **197** in a high yield, which upon treatment with hydrazine hydrate in ethanol at reflux temperature furnished the 3,4'-bipyrazole derivative **198** in 80% yield (Scheme **55**) [88].

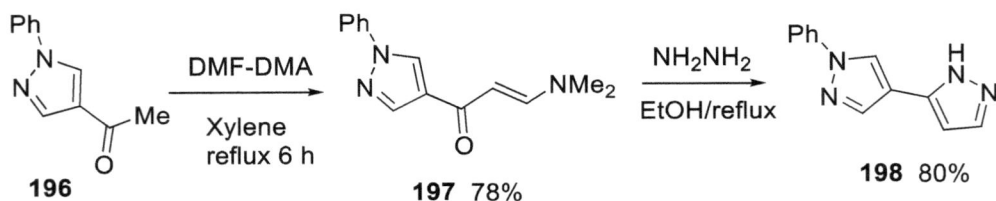

Scheme (55). Synthesis of the 3,4'-bipyrazole derivative **198**.

The reaction of the pyrazolyl-enaminone **199** with hydrazine or phenylhydrazine provided the 3,4'-bipyrazole derivatives **200** in high yields (Scheme **56**) [89].

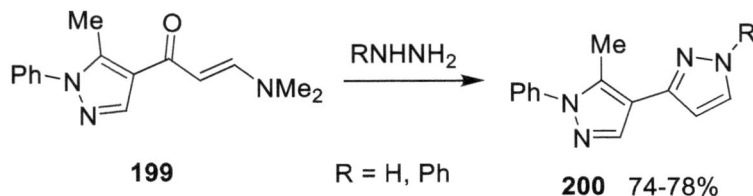

Scheme (56). Synthesis of the 3,4'-bipyrazole derivatives **200**.

The reaction of the pyrazolyl-enamine aldehydes **201** with hydrazine or phenylhydrazine produced the 3,4'-bipyrazole derivatives **202** (Scheme **57**) [90].

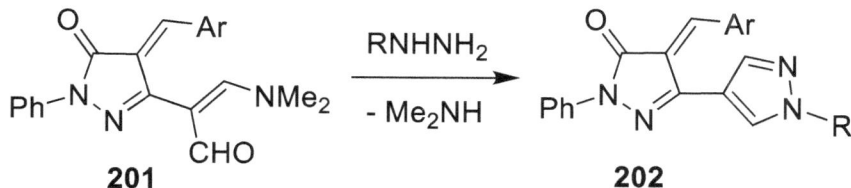

Scheme (57). Synthesis of the 3,4'-bipyrazole derivatives **202**.

The reaction of cyanoacetylantipyrine **203** with arylhydrazines produced the 5-amino-1',5'-dimethyl-3,4'-bipyrazol-3'-one derivatives **204**. Heating the latter amino-bipyrazole derivative **204** (R= H) with aromatic aldehydes in boiling ethanol using piperidine as a basic catalyst afforded the corresponding 5-amino-4-arylidene-3,4`-bipyrazol-3'-one derivatives **205** in good yields (Scheme **58**) [91 - 93].

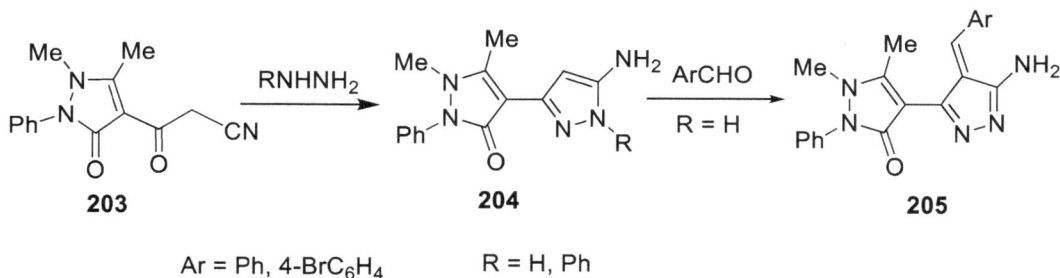

Ar = Ph, 4-BrC$_6$H$_4$ R = H, Ph

Scheme (58). Synthesis of the 3,4'-bipyrazole derivatives **205**.

The reaction of *N*-(benzothiazol-2-yl)-2-cyanoacetamide **206** with 1,3-diphenylpyrazole-4-carboxaldehyde **36** in ethanol, using sodium hydroxide (10%) as a catalyst for Knoevenagel condensation, afforded 2-cyano-3-(pyraz-l-4-yl)acrylamide derivative **207**. 5-Amino-1`,3`-diphenyl-3,4`-bipyrazole-4-carboxamide **208** was obtained from the addition of hydrazine hydrate to the activated double bond of compound **207** in boiling ethanol (Scheme **59**) [94].

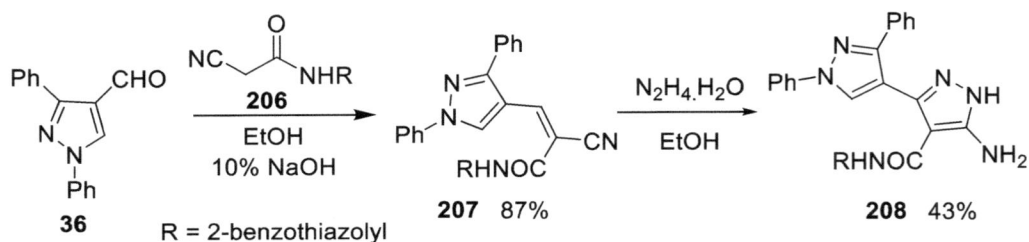

Scheme (59). Synthesis of the 3,4'-bipyrazole derivative **208**.

The 3,4`-bipyrazole derivative **210** was isolated from the reaction of 3-(diformylmethyl)-4-nitropyrazole **209** with hydrazine hydrochloride in a basic aqueous solution. Alternatively, the same 3,4`-bipyrazole derivative **210** was obtained more efficiently from the reaction of the perchlorate trimethinium salt **211** with hydrazine under similar reaction condition using a double amount of the base (Scheme **60**) [95].

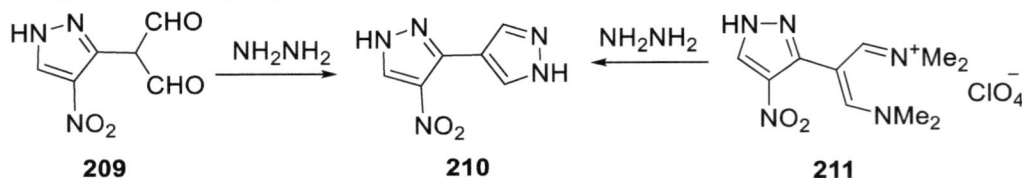

Scheme (60). Synthesis of the 3,4'-bipyrazole derivative **210**.

The 1,3-cycloaddition reaction of 1,4-diphenyl-1,3-butadiyne **212** with 2-diazopropane **213** furnished a mixture of two isomeric products; 4-phenylethynylpyrazole **214** and 5-phenylethynylpyrazole **215**. Treatment of the 4-phenylethynylpyrazole derivative **214** with 2-diazopropane **213** afforded the 3,4'-bipyrazole derivative **216** in a good yield (Scheme **61**) [96, 97].

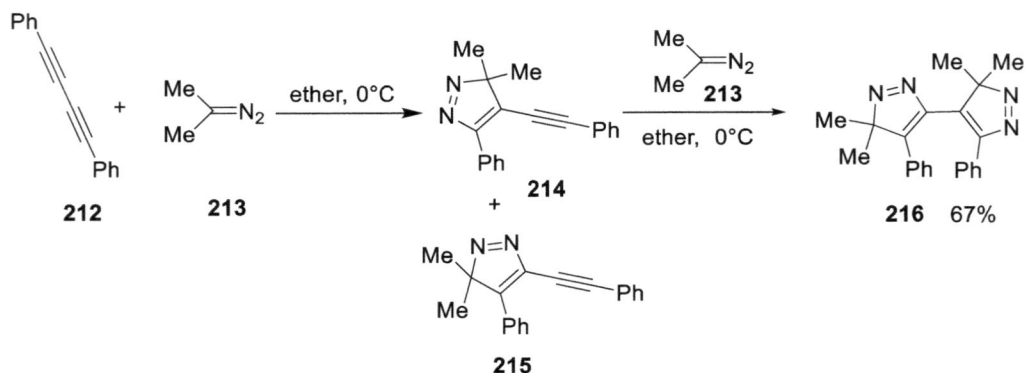

Scheme (61). Synthesis of the 3,4'-bipyrazole derivative **216**.

2.4. From C-C Cross Coupling Between Two Pyrazole Units

5-Pyrazolylboronic ester **218** underwent Suzuki cross-coupling reaction with 4-iodo-1-[2-(trimethylsilyl)ethoxy]methyl pyrazole **217** in the presence of $Pd(PPh_3)_4$ and $NaHCO_3$ in dimethoxyethane (DME)/water at reflux temperature, to afford the 3,4'-bipyrazole derivative **219** in a moderate yield. Deprotection of the latter compound **219** using *n*-Bu$_4$NF and ethylenediamine in THF at reflux condition led to the removal of only one SEM group of the 3,4'-bipyrazole **219** to give the mono-substituted 3,4'-bipyrazole **220** in a reasonable yield (Scheme **62**) [98].

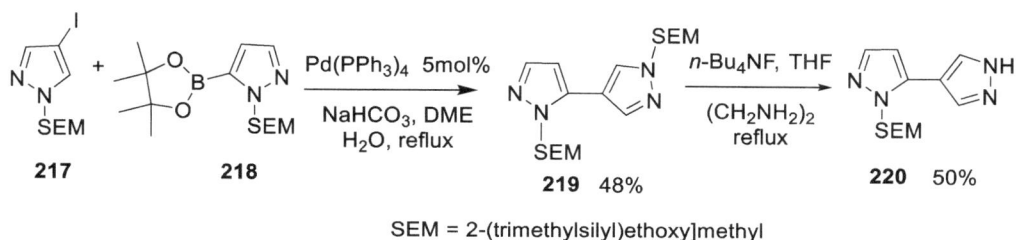

SEM = 2-(trimethylsilyl)ethoxy]methyl

Scheme (62). Synthesis of the 3,4'-bipyrazole derivatives **219** and **220**.

The 3,4'-bipyrazole-3,4-dicarboxylate ester **223** was synthesized in a reasonable yield by Suzuki cross-coupling reaction of the pyrazolyl-triflate derivative **222** with 4-pyrazolylboronic acid **221** catalysed by $Pd(PPh_3)_4$ (5 mol%) in anhydrous dimethoxyethane using Na_2CO_3 as a base (Scheme **63**) [99]. The ^1H NMR spectral data of compound **223** in $CDCl_3$ presented the following signals: δ 0.85 (6H, d, J = 6.6 Hz), 1.49 (1H, septet, J = 6.6 Hz), 1.67 (2H, dt, J = 7.8, 6.6 Hz), 3.72 (3H, s), 3.89 (3H, s), 3.97 (3H, s), 4.15– 4.19 (2H, m), 7.67 (1H, s), 7.94 (1H, s), and its ^{13}C NMR showed thirteen carbon-peaks at δ 22.6, 26.1, 30.0, 39.2, 49.2, 52.3, 52.7, 114.2, 131.6, 137.5, 142.2, 162.3, 163.7 ppm.

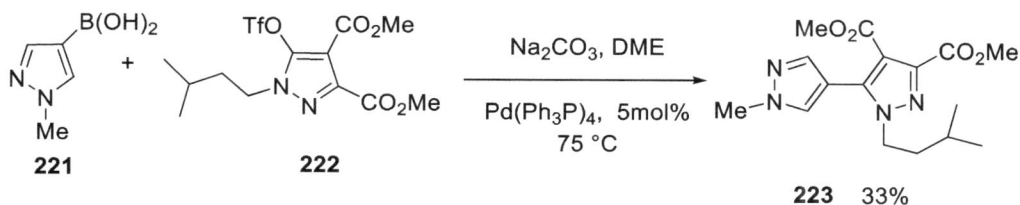

Scheme (63). Synthesis of the 3,4'-bipyrazole derivative **223**.

Reaction of nitropyrazloyl chloride **225** with 3-methylamino-pyrazole **224** in refluxing chloroform using triethylamine as a base gave the pyrazolylamide derivative **226**. The nitro function of compound **226** was reduced to the corresponding amine by the action of iron in acetic acid and then converted into the diazonium salt **227** as depicted in Scheme (**64**). Treatment of **227** with CuSO₄ in the presence of sodium chloride and ascorbic acid led to the formation of the spiroheterocyclic derivative **228**. Treatment of compound **228** with KOH in EtOH at ambient temperature provided the 3,4'-bipyrazole derivative **229** in a good yield (Scheme **64**) [100].

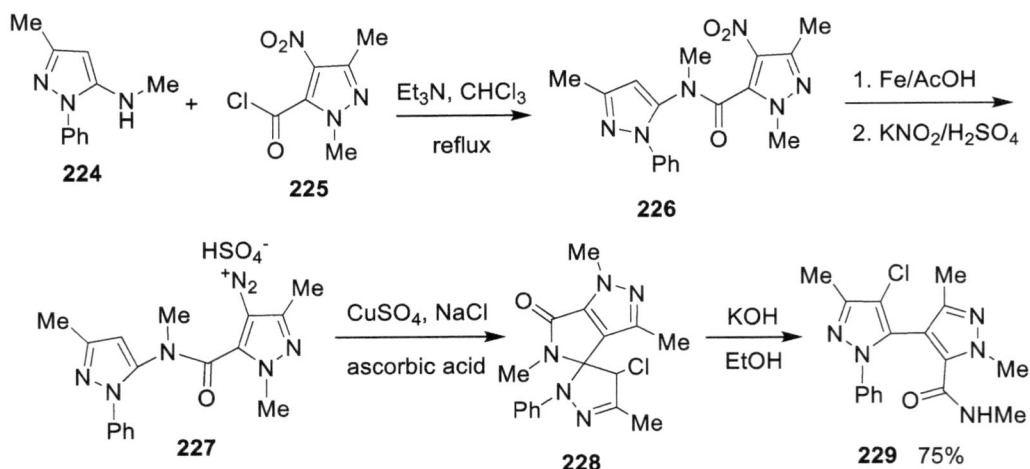

Scheme (64). Synthesis of the 3,4'-bipyrazole derivative **229**.

CONCLUSION

Synthesis of the fully aromatic 3,4`-bipyrazoles and their partially aromatic ones (pyrazolylpyrazolines) or 3,4`-bipyrazolines was reported extensively. Among the synthetic employed platform, cyclocondensation of pyrazole-based chalcones or 1,3-dicarbonyl functions with hydrazines as well as the 1,3-dipolar cycloaddition of pyrazolylhydrazones or nitrilimines in addition to the metal catalyzed cross-coupling reactions were comprehensively studied. Further applications 3,4`-bipyrazoles may be investigated by medicinal and chemistry researchers as well as those who are interested in the material science field.

REFERENCES

[1] Nawaz, F.; Alam, O.; Perwez, A.; Rizvi, M.A.; Naim, M.J.; Siddiqui, N.; Pottoo, F.H.; Jha, M. 3'-(-
     -(Benzyloxy)phenyl)-1'-phenyl-5-(heteroaryl/aryl)-3,4-dihydro-1'H,2H-[3,4'-bi-
     yrazole]-2-carboxamides as EGFR kinase inhibitors: Synthesis, anticancer evaluation, and molecular
     docking studies. *Arch. Pharm. (Weinheim), 2020, 353*(4), 1900262.
     [http://dx.doi.org/10.1002/ardp.201900262]

[2] Badr, M.H.; Abd El Razik, H.A. 1,4-Disubstituted-5-hydroxy-3-methylpyrazoles and some derived
     ring systems as cytotoxic and DNA binding agents. Synthesis, *in vitro* biological evaluation and *in
     silico* ADME study. *Med. Chem. Res., 2018, 27*(2), 442-457.
     [http://dx.doi.org/10.1007/s00044-017-2071-y]

[3]   Gul, H.I.; Yamali, C.; Bulbuller, M.; Kirmizibayrak, P.B.; Gul, M.; Angeli, A.; Bua, S.; Supuran, C.T. Anticancer effects of new dibenzenesulfonamides by inducing apoptosis and autophagy pathways and their carbonic anhydrase inhibitory effects on hCA I, hCA II, hCA IX, hCA XII isoenzymes. *Bioorg. Chem.,* **2018**, *78,* 290-297.
        [http://dx.doi.org/10.1016/j.bioorg.2018.03.027] [PMID: 29621641]

[4]   Gajera, S.B.; Mehta, J.V.; Thakor, P.; Thakkar, V.R.; Chudasama, P.C.; Patel, J.S.; Patel, M.N. Half-sandwich iridium III complexes with pyrazole-substituted heterocyclic frameworks and their biological applications. *New J. Chem.,* **2016**, *40*(12), 9968-9980.
        [http://dx.doi.org/10.1039/C6NJ02153K]

[5]   Kanthecha, D.N.; Bhatt, B.S.; Patel, M.N.; Raval, D.B.; Thakkar, V.R.; Vaidya, F.U.; Pathak, C. Bipyrazole Based Novel Bimetallic μ-oxo Bridged Au(III) Complexes as Potent DNA Interacalative, Genotoxic, Anticancer, Antibacterial and Cytotoxic Agents. *J. Inorg. Organomet. Polym. Mater.,* **2020**, *30*(12), 5085-5099.
        [http://dx.doi.org/10.1007/s10904-020-01618-2]

[6]   Lunagariya, M.V.; Thakor, K.P.; Pursuwani, B.H.; Patel, M.N. Evolution of 1, 3, 5-trisubstituted bipyrazole scaffold based platinum(II) complexes as a biological active agent. *Nucleos Nucleot Nucleic Acids,* **2018**, *37*(8), 455-483.
        [http://dx.doi.org/10.1080/15257770.2018.1498510] [PMID: 30230996]

[7]   Siddiqui, Z.N.; Musthafa, T.N.; Ahmad, A.; Khan, A.U. Thermal solvent-free synthesis of novel pyrazolyl chalcones and pyrazolines as potential antimicrobial agents. *Bioorg. Med. Chem. Lett.,* **2011**, *21*(10), 2860-2865.
        [http://dx.doi.org/10.1016/j.bmcl.2011.03.080] [PMID: 21507638]

[8]   Dhinoja, V.; Karia, D.; Shah, A. Acid promoted one pot synthesis of some new coumarinyl 3,4′-bipyrazole and their *in vitro* antimicrobial evaluation. *Chem. Biol. Interact.,* **2014**, *4,* 232-245.

[9]   Parshad, M.; Kumar, D. Design, synthesis and characterization of isomeric 3,4′-bipyrazol-5′-ols and their antifungal activity. *Chem. Biol. Interact.,* **2014**, *4,* 100-110.

[10]   Kalaria, P.N.; Makawana, J.A.; Satasia, S.P.; Ravala, D.K.; Zhub, H-L. Design, synthesis and molecular docking of novel bipyrazolyl thiazolone scaffold as a new class of antibacterial agents. *MedChemComm,* **2014**, *5*(10), 1555-1562.
        [http://dx.doi.org/10.1039/C4MD00238E]

[11]   Pundeer, R.; Kiran, V.; Sharma, C.; Aneja, K.R.; Prakash, O. Synthesis and evaluation of antibacterial and antifungal activities of new (Z)-3-bromo-4-(1,3-diaryl-1H-pyrazol-4-yl) but-3-en-2-ones and 4-(--methyl-1-phenyl-1H-pyrazol-5-yl)-1,3-diaryl-1H-pyrazoles. *Med. Chem. Res.,* **2013**, *22*(10), 4715-4726.
        [http://dx.doi.org/10.1007/s00044-013-0480-0]

[12]   Karad, S.C.; Purohit, V.B.; Raval, D.K. Design, synthesis and characterization of fluoro substituted novel pyrazolylpyrazolines scaffold and their pharmacological screening. *Eur. J. Med. Chem.,* **2014**, *84,* 51-58.
        [http://dx.doi.org/10.1016/j.ejmech.2014.07.008] [PMID: 25016227]

[13]   Taher, A.T.; Mostafa Sarg, M.T.; El-Sayed Ali, N.R.; Hilmy Elnagdi, N. Design, synthesis, modeling studies and biological screening of novel pyrazole derivatives as potential analgesic and anti-inflammatory agents. *Bioorg. Chem.,* **2019**, *89,* 103023.
        [http://dx.doi.org/10.1016/j.bioorg.2019.103023] [PMID: 31185391]

[14]   Faidallah, H.M.; Rostom, S.A.F. Synthesis, Anti-Inflammatory Activity, and COX-1/2 Inhibition Profile of Some Novel Non-Acidic Polysubstituted Pyrazoles and Pyrano[2,3-c]pyrazoles. *Arch. Pharm. (Weinheim),* **2017**, *350*(5), 1700025.
        [http://dx.doi.org/10.1002/ardp.201700025] [PMID: 28370254]

[15]   Faidallah, H.M.; Rostom, S.A.; Khan, K.A. Synthesis and biological evaluation of pyrazole chalcones and derived bipyrazoles as anti-inflammatory and antioxidant agents. *Arch. Pharm. Res.,* **2015**, *38*(2),

203-215.
[http://dx.doi.org/10.1007/s12272-014-0392-7] [PMID: 24752861]

[16]   Harras, M.F.; Sabour, R.; Alkamali, O.M. Discovery of new non-acidic lonazolac analogues with COX-2 selectivity as potent anti-inflammatory agents. *MedChemComm,* **2019**, *10*(10), 1775-1788.
[http://dx.doi.org/10.1039/C9MD00228F] [PMID: 31803395]

[17]   Ragab, F.A.E.; Mohammed, E.I.; Abdel Jaleel, G.A.; Selim, A.A.M.A.E.; Nissan, Y.M. Synthesis of Hydroxybenzofuranyl-pyrazolyl and Hydroxyphenyl-pyrazolyl Chalcones and Their Corresponding Pyrazoline Derivatives as COX Inhibitors, Anti-inflammatory and Gastroprotective Agents. *Chem. Pharm. Bull. (Tokyo),* **2020**, *68*(8), 742-752.
[http://dx.doi.org/10.1248/cpb.c20-00193] [PMID: 32741915]

[18]   Kumar, P.; Chandak, N.; Kaushik, P.; Sharma, C.; Kaushik, D.; Aneja, K.R.; Sharma, P.K. Benzenesulfonamide bearing pyrazolylpyrazolines: synthesis and evaluation as anti-inflammatory–antimicrobial agents. *Med. Chem. Res.,* **2014**, *23*(2), 882-895.
[http://dx.doi.org/10.1007/s00044-013-0679-0]

[19]   Sharma, P.K.; Kumar, S.; Kumar, P.; Kaushik, P.; Kaushik, D.; Dhingra, Y.; Aneja, K.R. Synthesis and biological evaluation of some pyrazolylpyrazolines as anti-inflammatory-antimicrobial agents. *Eur. J. Med. Chem.,* **2010**, *45*(6), 2650-2655.
[http://dx.doi.org/10.1016/j.ejmech.2010.01.059] [PMID: 20171763]

[20]   Ali, S.A.; Awad, S.M.; Said, A.M.; Mahgoub, S.; Taha, H.; Ahmed, N.M. Design, synthesis, molecular modelling and biological evaluation of novel 3-(2-naphthyl)-1-phenyl-1H-pyrazole derivatives as potent antioxidants and 15-Lipoxygenase inhibitors. *J. Enzyme Inhib. Med. Chem.,* **2020**, *35*(1), 847-863.
[http://dx.doi.org/10.1080/14756366.2020.1742116] [PMID: 32216479]

[21]   Mogle, P.P.; Meshram, R.J.; Hese, S.V.; Kamble, R.D.; Kamble, S.S.; Gacche, R.N.; Dawane, B.S. Synthesis and molecular docking studies of a new series of bipyrazol-yl-thiazol-ylid-ne-hydrazinecarbothioamide derivatives as potential antitubercular agents. *MedChemComm,* **2016**, *7*(7), 1405-1420.
[http://dx.doi.org/10.1039/C6MD00085A]

[22]   Harikrishna, N.; Isloor, A.M.; Ananda, K.; Obaid, A.; Kun, H. Synthesis, and antitubercular and antimicrobial activity of 1′-(4-chlorophenyl) pyrazole containing 3,5-disubstituted pyrazoline derivatives. *New J. Chem.,* **2016**, *40*(1), 73-76.
[http://dx.doi.org/10.1039/C5NJ02237A]

[23]   Neha, S.; Nitin, K.; Devender, Y.M.; Pathak, D. Synthesis of pyrazole derivatives: a new therapeutic approach for antitubercular and anticancer activity. *J. Pharm. Res.,* **2013**, *12*(1), 5-14.
[http://dx.doi.org/10.18579/jpcrkc/2013/12/1/79117]

[24]   Thillainayagam, M.; Ramaiah, S.; Anbarasu, A. Molecular docking and dynamics studies on novel benzene sulfonamide substituted pyrazole-pyrazoline analogues as potent inhibitors of *Plasmodium falciparum* Histo aspartic protease. *J. Biomol. Struct. Dyn.,* **2020**, *38*(11), 3235-3245.
[http://dx.doi.org/10.1080/07391102.2019.1654923] [PMID: 31411122]

[25]   Khloya, P.; Ceruso, M.; Ram, S.; Supuran, C.T.; Sharma, P.K. Sulfonamide bearing pyrazolylpyrazolines as potent inhibitors of carbonic anhydrase isoforms I, II, IX and XII. *Bioorg. Med. Chem. Lett.,* **2015**, *25*(16), 3208-3212.
[http://dx.doi.org/10.1016/j.bmcl.2015.05.096] [PMID: 26105196]

[26]   Yukimasa, A.; Kozono, I.; Nakamura, K. Preparation of heterocyclic compounds and their salts with TrkA inhibitory effect, and pharmaceutical compositions containing them. *PCT Int. Appl,* , WO 2016021629, A1.

[27]   Allen, S.; Andrews, S.W.; Baer, B.; Crane, Z.; Liu, W.; Watson, D.J. Preparation of 1-((3S,4R)-4-(3-fluorophenyl)-1-(2-methoxyethyl)pyrrolidin-3-yl)-3-(4-methyl-3-(2-methyl-pyrimidin-5-yl)-1-phenyl-1H-pyrazol-5-yl)urea as a TRKA kinase inhibitor. *PCT Int. Appl.,* , WO 2015175788 A1.

[28] Allen, S.; Andrews, S.W.; Blake, J.F.; Condroski, K.R.; Haas, J.; Huang, L.; Jiang, Y.; Kercher, T. Pyrrolidinyl urea and pyrrolidinylthiourea compounds as trka kinase inhibitors. *PCT Int. Appl.,* , WO 2012158413 A2.

[29] Andrews, S.W.; Blake, J.F.; Brandhuber, B.J.; Kercher, T.; Winski, S.L. N-aryl-N'-pyrazolyl-urea, thiourea, guanidine and cyanoguanidine compounds as TrkA kinase inhibitors and their preparation. *PCT Int. Appl.,* , WO 2014078325 A1.

[30] Allen, S.; Andrews, S.W.; Blake, J.F.; Brandhuber, B.J.; Haas, J.; Jiang, Y.; Kercher, T.; Kolakowski, G.R.; Thomas, A.A.; Winski, S.L. Preparation of bicyclic urea, thiourea, guanidine and cyanoguanidine compounds as TrkA inhibitors that are useful for the treatment of pain. *PCT Int. Appl.,* , WO 2014078454, A1.

[31] Li, Y.L.; Zhuo, J.; Qian, D.Q.; Mei, S.; Cao, G.; Pan, Y.; Li, Q.; Jia, Z. Preparation of bipyrazole derivatives as JAK kinase inhibitors. *PCT Int. Appl.,* , WO 2014186706 A1.

[32] Thakor, K.P.; Lunagariya, M.V.; Bhatt, B.S.; Patel, M.N. Bipyrazole-based palladium (II) complexes as DNA intercalator and artificial metallonuclease. *Monatsh. Chem.,* **2019**, *150*(2), 233-245.
[http://dx.doi.org/10.1007/s00706-018-2316-6]

[33] Arrieta, A.; Carrillo, J.R.; Cossio, F.P.; Diaz-Ortiz, A.; Gómez-Escalonilla, M.J.; de la Hoz, A.; Moreno, A.; Langa, F. Efficient Tautomerization Hydrazone-Azomethine Imine under Microwave Irradiation. Synthesis of [4,3'] and [5,3']Bipyrazoles. *Tetrahedron,* **1998**, *54*(43), 13167-13180.
[http://dx.doi.org/10.1016/S0040-4020(98)00798-4]

[34] Diaz-Ortiz, A.; de la Hoz, A.; Langa, F. Microwave irradiation in solvent-free conditions: an eco-friendly methodology to prepare indazoles, pyrazolopyridines and bipyrazoles by cycloaddition reactions. *Green Chem.,* **2000**, *2*(4), 165-172.
[http://dx.doi.org/10.1039/b003752o]

[35] Carrillo, J.R.; Cossio, F.P.; Diaz-Ortiz, A.; Gomez-Escalonilla, M.J.; de la Hoz, A.; Moreno, A.; Lecea, B.; Moreno, A.; Prieto, P. A complete model for the prediction of ^{1}H- and ^{13}C-NMR chemical shifts and torsional angles in phenyl-substituted pyrazoles. *Tetrahedron,* **2001**, *57*(19), 4179-4187.
[http://dx.doi.org/10.1016/S0040-4020(01)00291-5]

[36] Bougrin, K.; Loupy, A.; Soufiaoui, M. Microwave-assisted solvent-free heterocyclic synthesis. *J. Photochem. Photobiol. Photochem. Rev.,* **2005**, *6*(2-3), 139-167.
[http://dx.doi.org/10.1016/j.jphotochemrev.2005.07.001]

[37] de la Cruz, P.; Diaz-Ortiz, A.; Garcia, J.J.; Gómez-Escalonilla, M.J.; de la Hoz, A.; Langa, F. Synthesis of new C60-donor dyads by reaction of pyrazolylhydrazones with [60]fullerene under microwave irradiation. *Tetrahedron Lett.,* **1999**, *40*(8), 1587-1590.
[http://dx.doi.org/10.1016/S0040-4039(98)02651-3]

[38] Farag, A.M.; Kheder, N.A.; Dawood, K.M.; El Defrawy, A.M. A Facile Access and Computational Studies of Some New 4,5'-Bipyrazole Derivatives. *Heterocycles,* **2017**, *94*, 1245-1256.
[http://dx.doi.org/10.3987/COM-17-13707]

[39] Oida, T.; Tanimoto, S.; Ikehira, H.; Okano, M. The Cycloaddition of N,N-Diethyl-1-3-butadienylamine with Some Diarylnitrilimines. *Bull. Chem. Soc. Jpn.,* **1983**, *56*(4), 1203-1205.
[http://dx.doi.org/10.1246/bcsj.56.1203]

[40] Ruccia, M.; Vivona, N.; Cusmano, G. Addition reactions of heterocycles. VI. Reactions of 1,2-dimethylpyrrole and 1-methyl-2-carbomethoxypyrrole with nitrilimines. *J. Heterocycl. Chem.,* **1978**, *15*(2), 293-296.
[http://dx.doi.org/10.1002/jhet.5570150222]

[41] Zala, M.; Vora, J.J.; Patel, H.B. Synthesis, Characterization, and Comparative Study of Some Heterocyclic Compounds Containing Isoniazid and Nicotinic Acid Hydrazide Moieties. *Russ. J. Org. Chem.,* **2020**, *56*(10), 1795-1800.
[http://dx.doi.org/10.1134/S1070428020100218]

[42] Raut, D.G.; Lawand, A.S.; Kadu, V.D.; Hublikar, M.G.; Patil, S.B.; Bhosale, D.G.; Bhosale, R.B. Synthesis of Asymmetric Thiazolyl Pyrazolines as a Potential Antioxidant and Anti-Inflammatory Agents *Polycycl. Arom. Comp.,* **2020**. Ahead of Print [http://dx.doi.org/10.1080/10406638.2020.1716028]

[43] Halnor, V.B.; Joshi, N.S.; Karale, B.K.; Gill, C.H. Synthesis, characterization of some 1-(2-hydro-y-phenyl)-3-(l-phenyl-3-thiophen-2-yl-1H-pyrazol-4-yl)popenone. *Heterocycl. Commun.,* **2005**, *11,* 167-172.

[44] Kiran Kumar, H.; Yathirajan, H.S.; Manju, N.; Kalluraya, B.; Rathore, R.S.; Glidewell, C. Conversion of substituted 5-aryloxypyrazolecarbaldehydes into reduced 3,4'-bipyrazoles: synthesis and characterization, and the structures of four precursors and two products, and their supramolecular assembly in zero, one and two dimensions. *Acta Crystallogr. C Struct. Chem.,* **2019**, *75*(Pt 6), 768-776. [http://dx.doi.org/10.1107/S2053229619006752] [PMID: 31166931]

[45] Cuartas, V.; Insuasty, B.; Cobo, J.; Glidewell, C. Reduced 3,4'-bipyrazoles from a simple pyrazole precursor: synthetic sequence, molecular structures and supramolecular assembly. *Acta Crystallogr. C Struct. Chem.,* **2017**, *73*(Pt 10), 784-790. [http://dx.doi.org/10.1107/S205322961701302X] [PMID: 28978784]

[46] Faidallah, H.M.; Khan, K.A.; Asiri, A.M.; Zayed, M.E. Synthesis and biological evaluation of some new bipyrazolylbenzenesulfonamides as possible antimicrobial and chemotherapeutic agents. *ChemInform,* **2012**, *51B*(48), 1114-1122. [http://dx.doi.org/10.1002/chin.201248118]

[47] Gomha, S.M.; Salah, T.A.; Abdelhamid, A.O. Synthesis, characterization, and pharmacological evaluation of some novel thiadiazoles and thiazoles incorporating pyrazole moiety as anticancer agents. *Monatsh. Chem.,* **2015**, *146*(1), 149-158. [http://dx.doi.org/10.1007/s00706-014-1303-9]

[48] Reddy, P.N.; Reddy, Y.T.; Kumar, V.N.; Rajitha, B. Synthesis of new hetero aroyl chromen-4-ones. *Heterocycl. Commun.,* **2005**, *11*(3-4), 235-240. [http://dx.doi.org/10.1515/HC.2005.11.3-4.235]

[49] Siddiqui, Z.N.; Asad, M.; Praveen, S. Synthesis and biological activity of heterocycles from chalcone. *Med. Chem. Res.,* **2008**, *17*(2-7), 318-325. [http://dx.doi.org/10.1007/s00044-007-9067-y]

[50] Pawar, S.B.; Dalvi, N.R.; Karale, B.K.; Gill, C.H. Synthesis of some 4-(2-hydroxy phenyl)-6-(1, 3-diphenyl-1H-pyrazol-4-yl) pyrimidine-2 (1H)-thiones and 2 (5-(1, 3-diphenyl-1H-pyrazol-4--l)-1H-pyrazol-3-yl) phenols. *Indian J. Heterocycl. Chem.,* **2005**, *15,* 197-198.

[51] Levai, A.; Silva, A.M.S.; Pinto, D.C.G.A.; Cavaleiro, J.A.S.; Alkorta, I.; Elguero, J.; Jekö, J. Synthesis of Pyrazolyl-2-pyrazolines by Treatment of 3-(3-Aryl-3-oxopropenyl)-chromen-4-ones with Hydrazine and Their Oxidation to Bis(pyrazoles). *Eur. J. Org. Chem.,* **2004**, *2004*(22), 4672-4679. [http://dx.doi.org/10.1002/ejoc.200400465]

[52] Shelke, S.N.; Dalvi, N.R.; Gill, C.H.; Karale, B.K. Synthesis of Various Heterocycles from 3-(Naphthylene-3-yl)-1H-pyrazol-4-carbaldehyde. *Asian J. Chem.,* **2007**, *19,* 5068.

[53] Rashiid, S.; Khan, M.A.; Moazzam, M. Hetarylpyrazoles. Part VII. Some derivatives of 3,4'-bipyrazolyls. *J. Pure Appl. Sci.,* **1999**, *18,* 93-98.

[54] Narwade, S.K.; Kale, S.B.; Karale, B.K. Synthesis, and antimicrobial activities of some fluorinated substituted pyrazoles. *Indian J. Heterocycl. Chem.,* **2007**, *16*(3), 275-278.

[55] El-Shekeil, A.G.; Babagi, A.S.; Hassan, M.A.; Shiba, S.A. Synthesis and reactions of 1-aryl-3-(5-chloro-1,3-diphenyl-4-pyrazolyl)-2-propene-1-ones and their dibromides. *Proc. Pak. Acad. Sci.,* **1988**, *25,* 25-34.

[56] Chovatia, P.T.; Akabari, J.D.; Kachhadia, P.K.; Zalavadia, P.D.; Joshi, H.S. Synthesis and selective

antitubercular and antimicrobial inhibitory activity of 1-acetyl-3,5-diphenyl-4,5-dihydro-(1H)-pyrazole derivatives. *J. Serb. Chem. Soc.,* **2006**, *71*(7), 713-720.
[http://dx.doi.org/10.2298/JSC0607713C]

[57]   Bratenko, M.K.; Chornous, V.A.; Vovk, M.V. 4-Functionally Substituted 3-Heterylpyrazoles: IV. 1-Phenyl-3-aryl(heteryl)-5-(4-pyrazolyl)-2-pyrazolines. *Russ. J. Org. Chem.,* **2001**, *37*(4), 556-559.
[http://dx.doi.org/10.1023/A:1012442205046]

[58]   Mo, X-X.; Xie, Z-F.; Hui, Y-H.; Hu, J.; Liu, F-M. Synthesis and fluorescent properties of bipyrazoline compounds. *Yingyong Huaxue,* **2007**, *24*, 765-769.

[59]   Halnor, V.B.; Joshi, N.S.; Karale, B.K.; Gill, C.H. Synthesis and biological activities of some pyrazolines. *Indian J. Heterocycl. Chem.,* **2005**, *14*(4), 371-372.

[60]   El-Khawas, S.M.; Farghaly, A.M.; Chaaban, I.; Fahmy, S.M. The Difference in the Behaviour of Hydrazine and p-Substituted Phenylhydrazines on Various 4-(3-Aryl-3-Oxopropenyl) Antipyrines. *J. Chin. Chem. Soc. (Taipei),* **1990**, *37*(6), 605-609.
[http://dx.doi.org/10.1002/jccs.199000083]

[61]   El-Sakka, I.A.; Kandil, A.; Elmoghayar, M.H. Reactions with 3-Pyrazolin-5-ones: Synthesis of some 4-Substituted 2,3-Dimethyl-1-phenyl-3-pyrazolin-5-ones. *Arch. Pharm. (Weinheim),* **1983**, *316*(1), 76-82.
[http://dx.doi.org/10.1002/ardp.19833160115]

[62]   Sahoo, U.; Dhanya, B.; Seth, A.K.; Sen, A.K.; Kumar, S.; Yadav, Y.C.; Ghelani, T.K.; Chawla, R. Microwave assisted synthesis and characterization of certain novel bipyrazole derivatives and their antimicrobial activities. *Internat. J. Pharm. Res.,* **2010**, *2*, 82-87.

[63]   Kamani, K.A.; Patel, K.D. Synthesis, Characterization and biological evaluation of coumarin-pyrazole-pyrazoline hybrids. *World J. Pharm. Res.,* **2017**, *6*, 939-953.

[64]   Ashoka, D.; Ganesha, A.; Ravia, S.; Lakshmia, B.V.; Ramesh, B. One pot multicomponent synthesis of 3′,5-diaryl-1′-phenyl-3,4-dihydro-1′H,2H-3,4′-bipyrazoles and their antimicrobial activity. *Russ. J. Gen. Chem.,* **2014**, *84*(11), 2248-2256.
[http://dx.doi.org/10.1134/S1070363214110346]

[65]   Obydennov, D.L.; Khammatova, L.R.; Eltsov, O.S.; Sosnovskikh, V.Y. A chemo- and regiocontrolled approach to bipyrazoles and pyridones *via* the reaction of ethyl 5-acyl-4-pyrone-2-carboxylates with hydrazines. *Org. Biomol. Chem.,* **2018**, *16*(10), 1692-1707.
[http://dx.doi.org/10.1039/C7OB02725G] [PMID: 29451283]

[66]   Khan, M.A.; Cosenza, A.G. Bihetaryls. 4. A 3,4`-bipyrazolyl from hydrogen and 4H-pyrano[2,-c]pyrazol-4-one. *Afinidad,* **1988**, *45*(414), 173-174.

[67]   Kosley, R.W.; MacDonald, D.; Sher, R. Dipyrazoles as central nervous system agents. **2006**, WO 2006101903 A1.

[68]   Gelin, S.; Chantegrel, B.; Nadi, A.I. Synthesis of 4-(acylacetyl)-1-phenyl-2-pyrazolin-5-ones from 3-acyl-2H-pyran-2, 4 (3H)-diones. Their synthetic applications to functionalized 4-oxopyrano [2, 3-c] pyrazole derivatives. *J. Org. Chem.,* **1983**, *48*(22), 4078-4082.
[http://dx.doi.org/10.1021/jo00170a041]

[69]   Singh, S.P.; Prakash, O.; Vaid, R.K. Synthesis of some 2-(3′,5′-dimethylpyrazol-1′-yl)-4-m-thyl-6-substituted-quinolines and their 4′-substituted analogs. *Indian J. Chem.,* **1986**, *25B*, 945-950.

[70]   Afridi, A.S.; Katritzky, A.R.; Ramsden, C.A. Preparation of NN′-linked bi (heteroaryls) from dehydroacetic acid and 2, 6-dimethyl-4-pyrone. *J. Chem. Soc., Perkin Trans. 1,* **1977**, (12), 1428-1436.
[http://dx.doi.org/10.1039/P19770001428]

[71]   Singh, S.P.; Naithani, R.; Aggarwal, R.; Prakash, O. C-C Bond Cleavage Studies in Bipyrazoles: A Convenient Synthesis of Pyrazolo-5-ols. *Synth. Commun.,* **2005**, *35*(4), 611-619.
[http://dx.doi.org/10.1081/SCC-200049811]

[72]   Singh, S.P.; Tarar, L.S.; Kumar, D. Reaction of 1-[5-Hydroxy-3-methyl-1-(2-thiazolyl)-4-p-razolyl]-1,3-butanediones with phenyl and heterocyclic hydrazines: a convenient synthesis of 4, 5-bipyrazoles. *Indian J. Heterocycl. Chem.,* **1993**, *3*, 5-8.

[73]   Djerrari, B.; Essassi, E.; Fifani, J. Etude de la réaction d'hydrazinolyse de l'acide déshydroacétique. *Bull. Soc. Chim. Fr.,* **1991**, (4), 521-542.

[74]   Kumar, D.; Singh, S.P. The structure of the products resulting from dehydroacetic acid and hydrazines. *J. Indian Chem. Soc.,* **2006**, *83*(5), 419-426.

[75]   Gantos, A.; De March, P.; Moreno-Manas, M.; Pla, A.; Sanchez-Ferrando, F.; Virgili, A. Synthesis of pyrano [4, 3-c] pyrazol-4 (1H)-ones and-4 (2H)-ones from dehydroacetic acid. Homo-and eteronuclear selective NOE measurements for unambiguous structure assignment. *Bull. Chem. Soc. Jpn.,* **1987**, *60*(12), 4425-4431.
[http://dx.doi.org/10.1246/bcsj.60.4425]

[76]   Kumar, D.; Singh, S.P.; Martínez, A.; Fruchier, A.; Elguero, J.; Martínez-Ripoll, M.; Carrió, J.S.; Virgili, A. The structure of the compounds resulting from the reaction of arylhydrazines with dehydroacetic acid: an NMR and crystallographic study. *Tetrahedron,* **1995**, *51*(16), 4891-4906.
[http://dx.doi.org/10.1016/0040-4020(95)00172-5]

[77]   Elguero, J.; Marffnez, A.; Singh, S.P.; Grover, M.; Tatar, L.S.A. [1]H and [13]C Nmr study of the structure and tautomerism of 4-pyrazolylpyrazolinones. *J. Heterocycl. Chem.,* **1990**, *27*(4), 865-870.
[http://dx.doi.org/10.1002/jhet.5570270409]

[78]   Eiden, F.; Teupe, E.G. Pyridon-,Pyrazol-und Pyrimidin-Derivate aus 3,5-Diacyl-4-pyronen. *Arch. Pharm. (Weinheim),* **1979**, *312*(10), 863-872.
[http://dx.doi.org/10.1002/ardp.19793121013]

[79]   Singh, K.; Sharma, P.K. Synthesis, characterization and antimicrobial study of some benzenesulfonamide based bipyrazoles. *Int. J. Pharm. Pharm. Sci.,* **2014**, *6*(10), 345-351.

[80]   Mor, S.; Mohil, R.; Nagoria, S.; Kumar, A. Synthesis and antimicrobial evaluation of some 1-(--arylthiazol-2-yl)-1′-(aryl/heteroaryl)-3,3′-dimethyl-[4,5′-bi-1H-pyrazol]-5-ols. *J. Serb. Chem. Soc.,* **2017**, *82*(2), 127-139.
[http://dx.doi.org/10.2298/JSC160310002M]

[81]   Bendaas, A.; Hamdi, M.; Sellier, N. Synthesis of bipyrazoles and pyrazoloisoxazoles from 3-acetyl-4-hydroxy-6-methyl-2H-pyran-2-one. *J. Heterocycl. Chem.,* **1999**, *36*(5), 1291-1294.
[http://dx.doi.org/10.1002/jhet.5570360529]

[82]   Saidoun, A.; Boukenna, L.; Rachedi, Y.; Talhi, O.; Laichi, Y.; Lemouari, N.; Trari, M.; Bachari, K.; Silva, A.M. One-Pot Three-Component Synthesis of Bispyrazole-thiazole-pyran-2-one Heterocyclic Hybrids. *Synlett,* **2018**, *29*(13), 1776-1780.
[http://dx.doi.org/10.1055/s-0037-1610183]

[83]   Zuev, M.A.; Sukhanova, A.A.; Smola, A.G.; Prezent, M.A.; Proshin, A.N.; Baranin, S.V.; Bubnov, Y.N. Boron-chelate assisted synthesis of new bipyrazole derivatives. *Mendeleev Commun.,* **2018**, *28*(6), 612-614.
[http://dx.doi.org/10.1016/j.mencom.2018.11.016]

[84]   Ali, T.E.; Assiri, M.A.; Ibrahim, M.A.; Yahia, I.S. Nucleophilic Reactivity of a Novel 3-Chloro-3-(4,9-dimethoxy5-oxo-5H-furo[3,2-g]chromen-6-yl)prop-2-enal. *Russ. J. Org. Chem.,* **2020**, *56*(5), 845-855.
[http://dx.doi.org/10.1134/S1070428020050188]

[85]   Simionescu, C.; Comanita, E.; Vata, M.; Daranga, M. Syntheses of pyrazole monomers. *Angew. Makromol. Chem.,* **1977**, *62*(1), 135-144.
[http://dx.doi.org/10.1002/apmc.1977.050620111]

[86]   Okabe-Nakahara, F.; Masumoto, E.; Maruoka, H.; Yamagata, K. Synthesis of Novel Angular and Linear Fused [5-6-5] Heterocycles by the Reaction of Methyl Cyano-(3-cyano-4,5-dihydro--

(3H)-furanylidene) acetate with Hydrazines and Dimethylformamide Dimethyl Acetal. *Heterocycles,* **2018**, *96*(4), 664-676.
[http://dx.doi.org/10.3987/COM-17-13861]

[87]   Ali, Y.M.; Ismail, M.F.; Abu El-Azm, F.S.; Marzouk, M.I. Design, synthesis, and pharmacological assay of novel compounds based on pyridazine moiety as potential antitumor agents. *J. Heterocycl. Chem.,* **2019**, *56*(9), 2580-2591.
[http://dx.doi.org/10.1002/jhet.3662]

[88]   Hassaneen, H.M.; Abdelhamid, I.A. Acetylacetaldehyde Dimethyl Acetal as Versatile Precursors for the Synthesis of Arylazonicotinic Acid Derivatives: Green Multicomponent Syntheses of Bioactive Poly-Heteroaromatic Compounds. *J. Heterocycl. Chem.,* **2017**, *54*(2), 1048-1053.
[http://dx.doi.org/10.1002/jhet.2673]

[89]   Shaaban, M.R.; Eldebss, T.M.A.; Darweesh, A.F.; Farag, A.M. A convenient synthesis of pyrazole-substituted heterocycles. *J. Chem. Res.,* **2010**, *34*(1), 8-11.
[http://dx.doi.org/10.3184/030823409X12608188070442]

[90]   Awad, I.M. Studies in the Vilsmeier-Haack reaction, part VII: Synthesis and reaction of 3-methyl-1-phenyl-4-acetyl hydrazono 2-pyrazoline-5-one(-5-thione). *Monatsh. Chem.,* **1990**, *121*(12), 1023-1030.
[http://dx.doi.org/10.1007/BF00809252]

[91]   Elmaati, T.M.A.; El-Taweel, F.M. Routes to Pyrazolo [3,4-e][1,4] thiazepine, Pyrazolo [1, 5-a] pyrimidine and Pyrazole Derivatives. *J. Chin. Chem. Soc. (Taipei),* **2003**, *50*(3A), 413-418.
[http://dx.doi.org/10.1002/jccs.200300063]

[92]   Sofan, M.A.; El-Taweel, F.M.A.; Abu Elmaati, T.M.; Elagamey, A.A. Reactions with heterocyclic amidines: synthesis of pyrazolo[3,4-e]thiazepine, pyrazolo[1,5-c] 1,2,4-triazine and pyrazolo [1,5-a] pyrimidine derivatives. *Pharmazie,* **1994**, *49*, 482-486.
[http://dx.doi.org/10.1002/chin.199445161]

[93]   El-Agamey, A.A.; El-Sakka, I.; El-Shahat, Z.; Elnagdi, M.H. Activated Nitriles in Heterocyclic Synthesis: Studies on the Chemistry of Antipyrin-4-ylacetonitrile. *Arch. Pharm. (Weinheim),* **1984**, *317*(4), 289-239.
[http://dx.doi.org/10.1002/ardp.19843170402]

[94]   Bondock, S.; Fadaly, W.; Metwally, M.A. Synthesis and antimicrobial activity of some new thiazole, thiophene and pyrazole derivatives containing benzothiazole moiety. *Eur. J. Med. Chem.,* **2010**, *45*(9), 3692-3701.
[http://dx.doi.org/10.1016/j.ejmech.2010.05.018] [PMID: 20605657]

[95]   Shevelev, S.A.; Dalinger, I.L.; Shkineva, T.K.; Ugrak, B.I. Nitropyrazoles. *Russ. Chem. Bull.,* **1993**, *42*(11), 1857-1861.
[http://dx.doi.org/10.1007/BF00699003]

[96]   Kuznetsov, M.A.; Dorofeeva, Yu.V.; Semenovskii, V.V.; Gindin, V.A.; Studennikov, A.N. Reaction of 2-diazopropane with diphenyldiacetylene and photolysis of the resulting 3H-pyrazoles. Synthesis of 3,3-dimethyl-1-phenyl-2-(phenylethynyl)cyclopropene - the first conjugated alkynylcyclopropene. *Zh. Obshch. Khim.,* **1991**, *61*, 2286-2300.

[97]   Kuznetsov, M.A.; Dorofeeva, Yu.V.; Semenovskii, V.V.; Gindin, V.A.; Studenikov, A.N. Synthesis of 3,3-dimethyl-1-phenyl-2-phenylethynylcyclopropene the first conjugated alkynyl-cyclopropene. *Tetrahedron,* **1992**, *48*(7), 1269-1280.
[http://dx.doi.org/10.1016/S0040-4020(01)90789-6]

[98]   Gérard, A-L.; Bouillon, A.; Mahatsekake, C.; Collot, V.; Raulta, S. Efficient and simple synthesis of 3-aryl-1H-pyrazoles. *Tetrahedron Lett.,* **2006**, *47*(27), 4665-4669.
[http://dx.doi.org/10.1016/j.tetlet.2006.04.125]

[99]   Dragovich, P.S.; Bertolini, T.M.; Ayida, B.K.; Li, L-S.; Murphy, D.E.; Ruebsam, F.; Sun, Z.; Zhou, Y. Regiospecific synthesis of 1,5-disubstituted-1H-pyrazoles containing differentiated 3,4-dicarboxylic

acid esters *via* Suzuki coupling of the corresponding 5-trifluoromethane sulfonates. *Tetrahedron,* **2007**, *63*(5), 1154-1166.
[http://dx.doi.org/10.1016/j.tet.2006.11.053]

[100] Maggio, B.; Daidone, G.; Raffa, D.; Plescia, S.; Bombieri, G.; Meneghetti, F. Non-classical pschorr and sandmeyer reactions in pyrazole series. *Helv. Chim. Acta,* **2005**, *88*(8), 2272-2281.
[http://dx.doi.org/10.1002/hlca.200590161]

CHAPTER 4

Chemistry of 4,4`-Bipyrazoles

Abstract: Synthesis of a huge number of 4,4`-bipyrazole derivatives was achieved employing various synthetic platforms. This chapter outlines all possible routes (such as cyclocondensation, 1,3-dipolar cycloaddition and dimerization reactions) towards the construction of the 4,4`-bipyrazole heterocycles.

Keywords: 1,3-dipolar cycloaddition, 4,4`-bipyrazoles, Cross-coupling, Cyclo-condensation, Hydrazonoyl halides.

1. INTRODUCTION

The 4,4`-bipyrazole ring skeletons can have the possible tautomeric forms that are constructed in Fig. (**1**). Synthesis of 4,4`-bipyrazoles was achieved through a number of synthetic routes as outlined in Fig. (**2**). The reported synthetic routes are as follows: 1) cyclocondensation of the activated 4-pyrazole ring having dicarbonyl functions with hydrazines; 2) cyclocondensation of tetraketones or bis-enals with hydrazines; 3) 1,3-dipolar cycloaddition of nitrilimines or diazo-methane with bis-olefines, and 4) dimerization of pyrazole ring *via* electrolysis or homocoupling reactions using palladium catalysts.

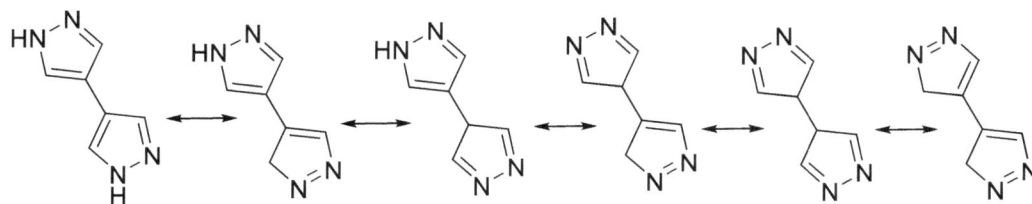

Fig. (1). The possible tautomeric forms of 4,4`-bipyrazole systems.

Kamal M. Dawood and Ashraf A. Abbas

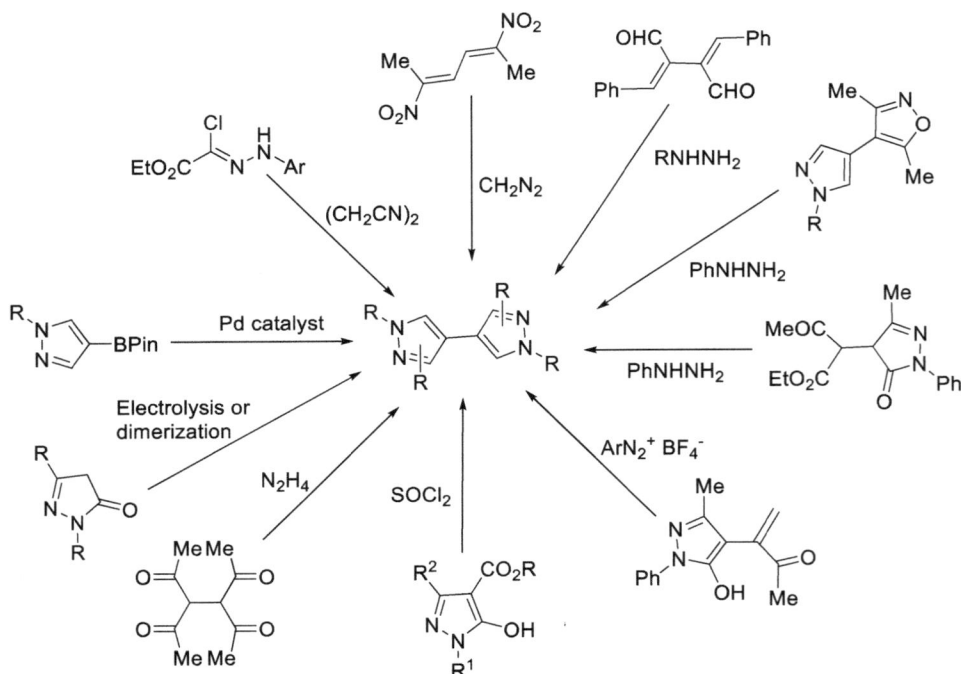

Fig. (2). The possible synthetic routes to 4,4`-bipyrazole systems.

4,4`-Bipyrazole derivatives were found to possess high biological potency and industrial applications. Some 4,4`-bipyrazole derivatives had a selective Janus kinase-1 (JAK1) inhibitory activity [1, 2]. Some 5,5'-dihydroxy-4,4'-bipyrazole derivatives were found to be useful for treatment of cerebral ischemia, heart diseases, gastrointestinal diseases, cancer, aging and inflammation, where they are effective in capturing the active oxygen and free radicals that are responsible for adult diseases [3 - 5]. Palladium(II) and platinum(II) complexes of 4,4'-bipyrazole were reported as potential anticancer agents [6], and the 4,4'-bipyrazol--Gadolonium(III) complexes were effective Paramagnetic Contrast Agent for clinical Magnetic Resonance Imaging (MRI) [7]. The nitrated 4,4`-bipyrazoles were classified as energetic and explosive materials [8, 9]. 4,4'-Bipyrazole systems were incorporated in the construction of several metal-organic frameworks (MOF). The MOF had promising diverse applications in drug delivery, gas separations, sensing, electrical conductivity, energy storage. and participated in forming porous coordination polymers with potential uses as solid sorbents, ion exchangers and heterogeneous catalysts [10 - 23].

2. SYNTHESIS OF 4,4`-BIPYRAZOLE DERIVATIVES

2.1. From Dimerization of Pyrazoles

Homocoupling of the pyrazolylboronic esters **1** and **3** catalyzed by Pd(PPh$_3$)$_4$ (5 mol%), in water solvent using Cs$_2$CO$_3$ as a base in the open air, led to the production of the symmetric 4,4'-bipyrazoles **2** and **4** in good yields, respectively (Scheme **1**) [11].

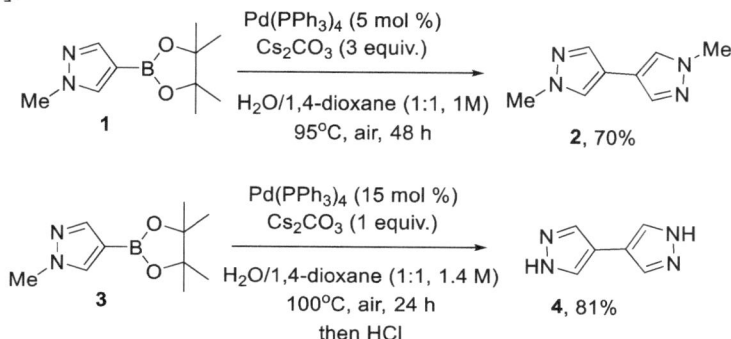

Scheme (1). Synthesis of the 3,4'-bipyrazole derivatives **2** and **4**.

Treatment of the pyrazolin-5-one **5** with Fe(ClO$_4$)$_3$ at ambient temperature led to its oxidative dimerization and formation of a diastereomeric mixture of 4,4'-bipyrazole-3,3'-diones **7** (*racemic,* 32% yield) and **8** (*meso,* 44% yield). The ^{1}H NMR spectral data of the *racemic* product **7** in CDCl$_3$ were as following: δ 1.60 (s, 6H, 2Me), 2.19 (s, 6H, 2Me), 7.18 (t, *J* = 7.3 Hz, 2H, ArH`s), 7.45–7.31 (m, 4H, ArH`s), 7.85 (d, *J* = 7.9 Hz, 4H, ArH`s); however the $^1$H NMR spectrum of the *meso*-compound **8** showed the following data: δ 1.73 (s, 6H, 2Me), 1.93 (s, 6H, 2Me), 7.22 (t, *J* = 7.3 Hz, 2H, ArH`s), 7.49–7.34 (m, 4H, ArH`s), 7.89 (d, *J* = 8.2 Hz, 4H, ArH`s). The reaction was supposed to took place *via* the pyrazolyl radical intermediate **6** as shown in Scheme (**2**) [24].

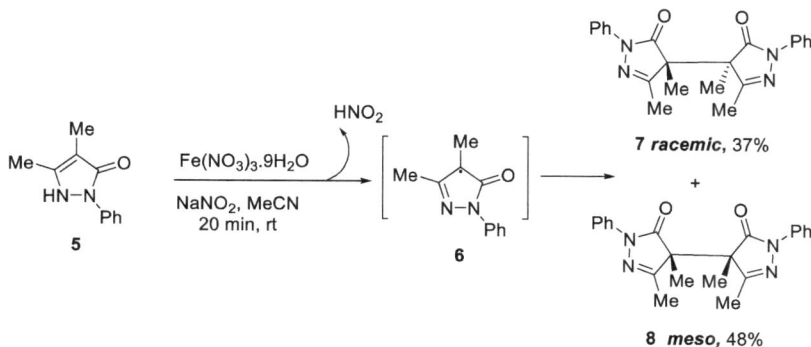

Scheme (2). Synthesis of 4,4`-bipyrazoles diastereomers **7** and **8**.

The diastereoselective reductive dimerization 4-tolylmethylene-3-phenylpyra-ol-5-one **9** was reported by Bruno *et. al*. The reaction took place *via* a radical pathway where a single electron transfer was generated *in situ* using 2-arylbenzimidazoline **10** as catalyst, to afford the corresponding 4,4`-bipyrazoline derivative **13** as shown in Scheme (3) [25, 26].

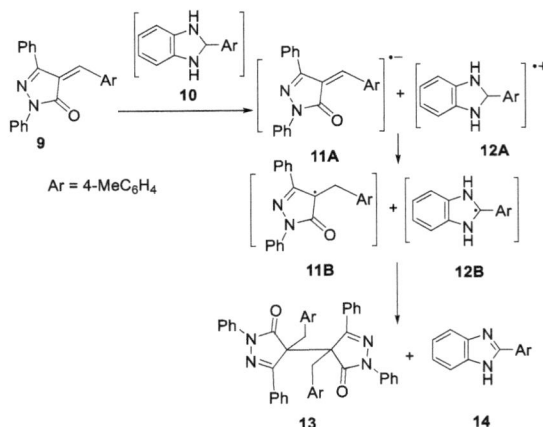

Scheme (3). Synthesis of the 3,4'-bipyrazole derivative **13**.

5-Hydroxypyrazole-4-carboxylate esters **15** were smoothly converted into the 5,5'-dioxo-4,4'-bipyrazole-4,4'-dicarboxylate esters **16**, in high yields when these were heated in the presence of thionyl chloride. The latter 4,4'-bipyrazole-4,4'-dicarboxylates **16** were hydrolysed in the presence of KOH then decarboxylated by the action of an aqueous hydrochloric acid leading to the formation of the 4,4'-bipyrazole-5,5'-diole derivatives **17** in good to excellent yields (Scheme 4). Mechanistically, 5-hydroxypyrazole-4-carboxylate **15** was transformed into the 5,5'-dioxo-4,4'-bipyrazole-4,4'-dicarboxylates **16** involving a redox cyclization of the di(pyrazolyl) sulphite intermediate **18** with concurrent elimination of sulfur monoxide [27].

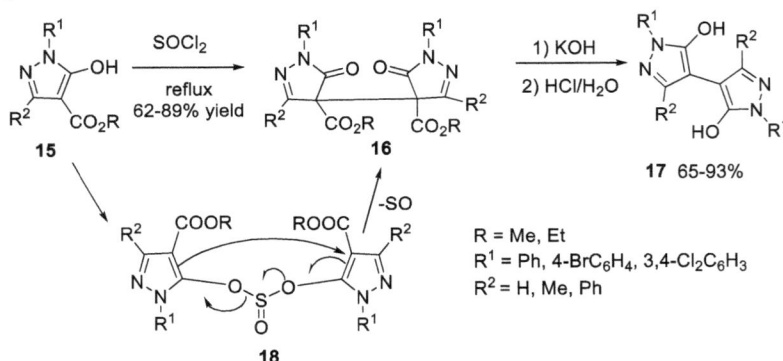

Scheme (4). Synthesis of the 4,4`-bipyrazol-5,5`-diol derivatives **17**.

Synthesis of the 4,4'-bipyrazol-5,5`-diol derivatives **21** was reported by Ueda *et al. via* dimerization of the pyrazolone **19** upon treatment with *N*-bromo-succinimide (NBS) or bromine in chloroform. The dimerization process took place *via the* initial formation of the bromopyrazole intermediate **20,** which then reacted with the pyrazolone **19** to give **21** *via* elimination of HBr molecule (Scheme **5**) [28].

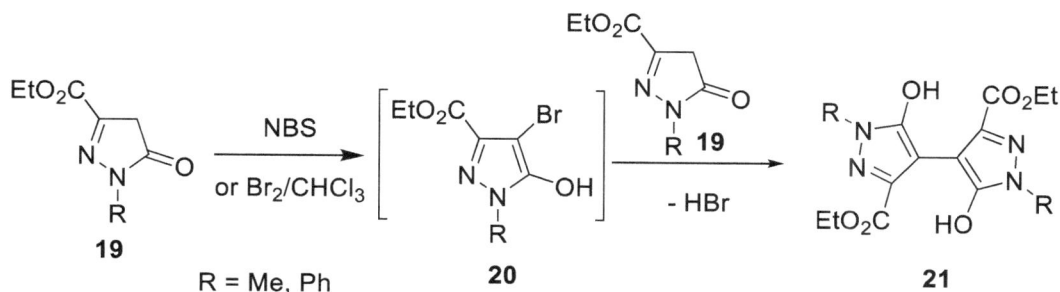

Scheme (5). Synthesis of the 3,4'-bipyrazole derivative **21**.

The pyrazolinone derivative **22** was also converted into the 4,4'-bipyrazol-5,5`-diol derivatives **24** *via* dimerization of the pyrazolyl radical **23**. Thus, when compound **22** was treated with 30% hydrogen peroxide, in the presence of selenium oxide in methanol at 0°C under a nitrogen atmosphere, it afforded the 4,4'-bipyrazol-5,5`-diol derivatives **24** (Scheme **6**) [3].

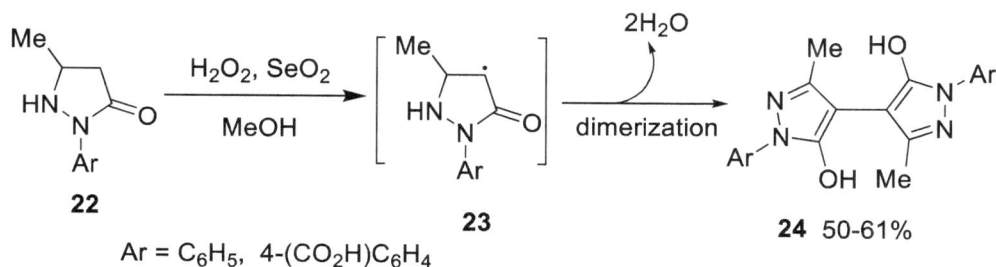

Scheme (6). Synthesis of the 3,4'-bipyrazole derivative **24**.

The 5-pyrazolone derivatives **25** underwent an electrochemical oxidative coupling when they were exposed to electric current with a density 3-8 mA/cm^2 in the presence of sodium halide in alcohol, as an electrolyte, and metal electrodes as the anode and cathode at -10~40 °C. The reaction generated the 4,4'-bipyrazole-5,5`-diol derivatives **26** in moderate yields (Scheme **7**) [29].

R= H, Ph, 4-ClC$_6$H$_4$, 4-MeC$_6$H$_4$, 4-MeOC$_6$H$_4$

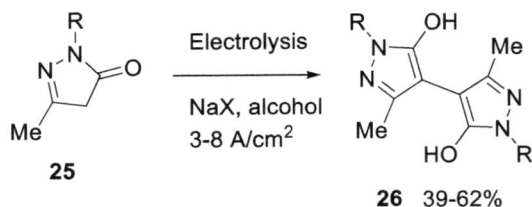

Scheme (7). Synthesis of the 4,4'-bipyrazol-5,5-diol derivatives **26**.

2.2. From 1,3-dipolar Cycloaddition

2,5-Dinitro-2,4-hexadiene **27** underwent smooth regioselective cycloaddition reaction upon treatment with diazomethane under the mild condition to provide 3,3'-dimethyl-3,3'-dinitro-4,4'-bipyrazole **29** in an excellent yield. The reaction proceeded *via* a double 1,3-dipolar cycloaddition, as shown in Scheme (**8**). Furthermore, the bipyrazolines **29** were readily aromatized by heating with sodium hydroxide in aqueous methanol to give 3,3'-dimethyl-4,4'-bipyrazole **30** in a high yield *via* loss of two molecules of nitrous acid from **29** (Scheme **8**) [30].

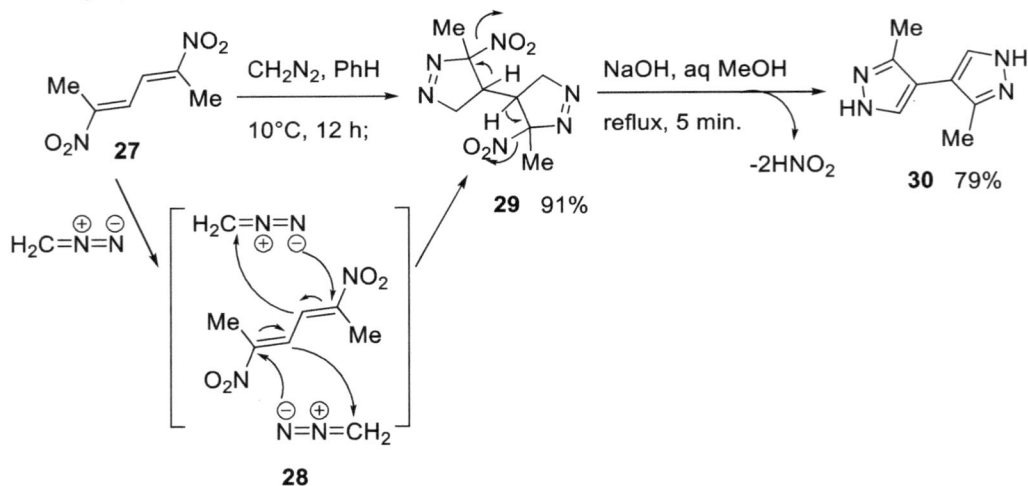

Scheme (8). Synthesis of 3,3'-dimethyl-4,4'-bipyrazoles **29** and **30**.

Treatment of succinonitrile **33** with the hydrazonoyl chloride **31**, in 1:2 molar ratio, in the presence of sodium ethoxide at ambient temperature, led to the formation of the 4,4'-bipyrazole derivative **34** in a good yield. The reaction proceeded *via* 1,3-dipolar cycloaddition of the nitrilimine intermediate **32** with the two nitrile functions of succinonitrile **33** (Scheme **9**) [31].

Scheme (9). Synthesis of the 3,4'-bipyrazole derivative **34**.

2.3. From Functionalized Pyrazoles

The 3,4-diacetylhexan-2,5-dione **36** was obtained from dimerization of acetylacetone **35** using iodine and NaOH. Treatment of the tetraketone **36** with hydrazine hydrate provided 3,3',5,5'-tetramethyl-4,4'-bipyrazole **37** in a good yield (Scheme **10**). Heating 4,4'-bipyrazole **37** with benzyl chlorides using tetrabuy-lammonium bromide, as a phase transfer catalyst, in toluene afforded the corresponding 1,1′-benzyl-4,4'-bipyrazole derivatives **38**. In addition, 5,5′-bi(2-bromoethylpyrazole) **39** was prepared in a low yield from a reaction of 4,4'-bipyrazole **37** with two equivalents of 1,2-dibromoethane under a liquid-liquid phase transfer catalysis (Scheme **10**) [7, 22, 23, 32].

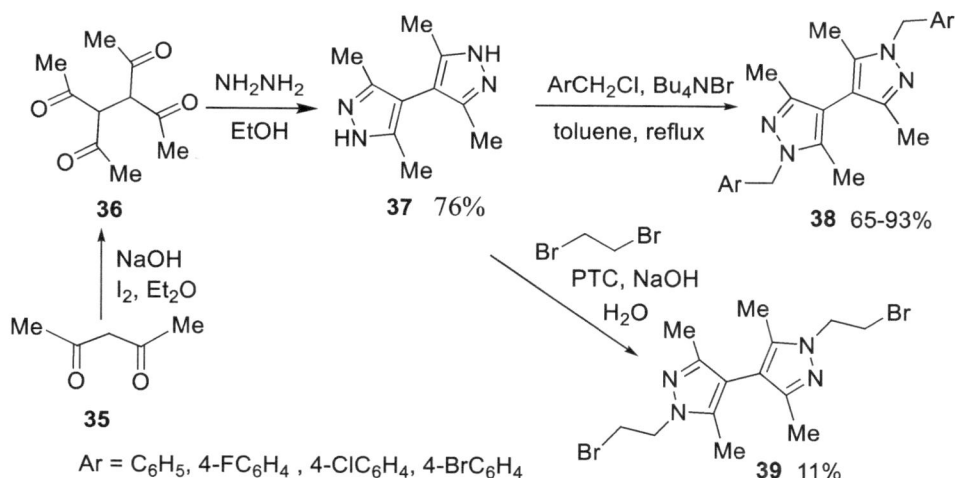

Scheme (10). Synthesis of the 3,4'-bipyrazole derivatives **38** and **39**.

Heating the pyrazolylisoxazole derivatives **40** with 1.2 equiv. of phenylhydrazine and 1 *M* HCl, in the presence of 0.5 equiv. of the molybdenum complex **41**, led to the production of the 4,4`-bipyrazole derivatives **44**. The reaction took place *via* Mo-based isoxazole ring opening with subsequent *in situ* hydrolysis of the β-imino-ketone **42** to give the 1,3-dicarbonyl compound **43**. Treatment of **43** with phenylhydrazine furnished the 4,4`-bipyrazole derivatives **44** (Scheme **11**) [33]. ^{1}H NMR of **44** (R = 4-F-C$_6$H$_4$-) showed signals at δ 2.38 (s, 3H), 2.43 (s, 3H), 7.12-7.22 (m, 2H), 7.40 (m, 1H), 7.49 (m, 4H), 7.74-7.66 (m, 2H), 7.79 (s, 1H), 7.89 (s, 1H); and its ^{13}C NMR spectrum a characteristic C-F carbon signal at δ 161.3 (d, *J* = 246.3 Hz) in addition to other fourteen C-signals at 147.2, 140.7, 139.4, 136.9, 136.6, 129.4, 128.0, 125.2, 124.9, 121.0, 116.4, 116.0, 111.4, 13.0, 12.1, in addition to its ^{19}F NMR that showed a characteristic F-signal at δ -115.7 ppm.

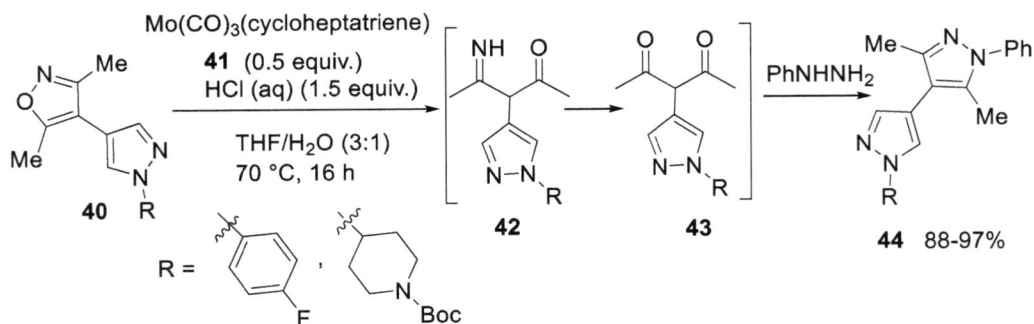

Scheme (11). Synthesis of the 4,4`-bipyrazole derivative **44**.

Nucleophilic 1,4-addition of the 2-pyrazolin-5-one derivatives **46** to the conjugated azoalkenes **45**, in the presence of Duolite® A1O$_2$ as a strong base anion exchanger resin, resulted in the formation of the 4,4'-bipyrazolin-3,3'-dione derivatives **50**. The reaction was supposed to proceed *via* the tautomeric intermediates **47**, **48**, and **49** with loss of methanol and carboxylic acid molecules (Scheme **12**) [34].

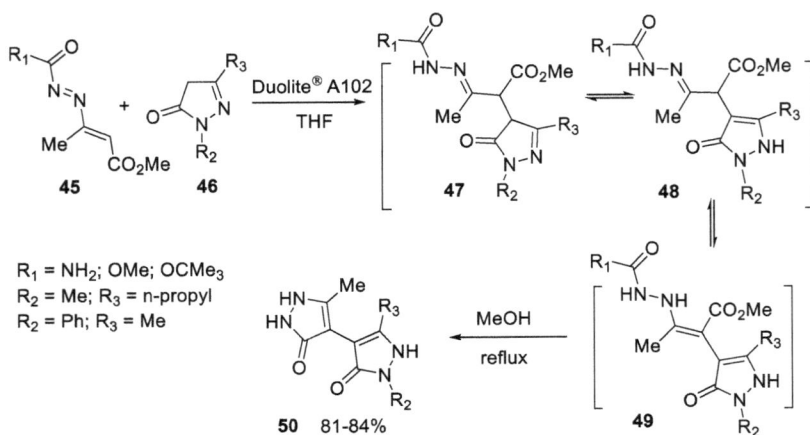

Scheme (12). Synthesis of the 3,4'-bipyrazole derivative **50**.

Treatment of the polyfunctionalized furan derivatives **51** with hydrazine generated the 4,4'-bipyrazole derivatives as outlined in Scheme **13**. Thus, heating a mixture of 4-acetyl-2-amino-5-methylfurans **51** with hydrazine in ethanol in the presence of trifluoroacetic acid (TFA) led to the formation of the 7-amino-4-5-dimethylfuro[3,4-d]pyridazine derivatives **52**. The latter fused heterocycles **52** were transformed into the corresponding 4,4'-bipyrazole derivatives **55** when these were heated with hydrazine in ethanol/acetonitrile mixed solvent (Scheme **13**). It was supposed that conversion of the furo[3,4-d]pyridazine **52** into 4,4`-bipyrazole **55** took place *via* ring opening to give the intermediate **53** and **54** then ring closure according to the mechanism outlined in Scheme **13** [35]. The ^{1}H NMR spectrum of compound **55** (X = NH) showed singlet peaks at δ 2.03 for methyl-protons and 5.56 and 8.5-10.5 for NH$_2$ and NH protons; and its ^{13}C NMR exhibited six C-peaks at δ 12.34, 80.47, 107.83, 142.84, 157.92, 170.61 ppm.

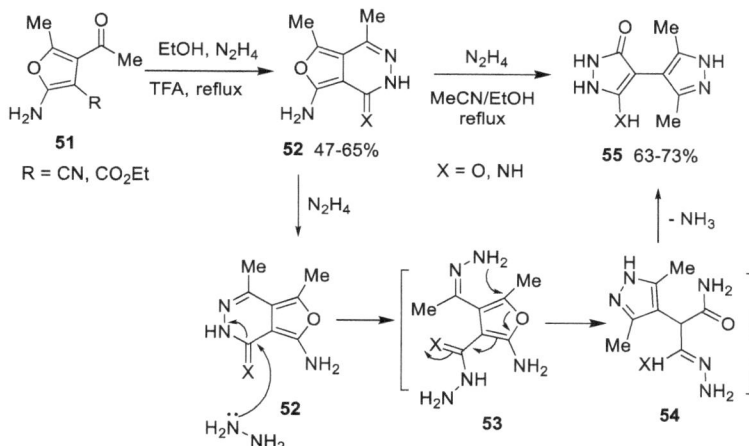

Scheme (13). Synthesis of the 3,4'-bipyrazole derivative **55**.

The reaction of the *bis*-aldehyde **56** with a number of hydrazine and its derivatives in refluxing ethanol provided the corresponding 4,4′-bipyrazole derivatives **57** in moderate yields (Scheme **14**) [36].

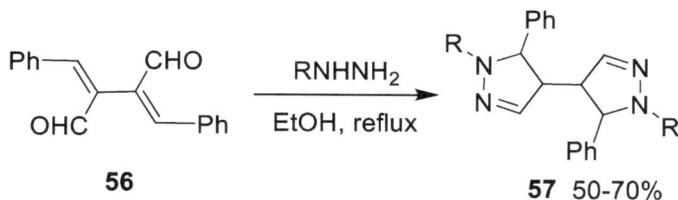

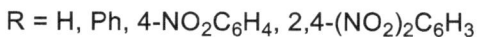

R = H, Ph, 4-NO$_2$C$_6$H$_4$, 2,4-(NO$_2$)$_2$C$_6$H$_3$

Scheme (14). Synthesis of the 3,4'-bipyrazole derivative **57**.

Synthesis of the 4,4'-bipyrazolylidene-3-one derivatives **61** from the cross-coupling reaction between 4-(α-acetylethylidene)pyrazole **58** and aryldiazonium fluoroborates **59** *via* the arylhydrazone intermediates **60** was reported by Lycka *et. al* (Scheme **15**) [37].

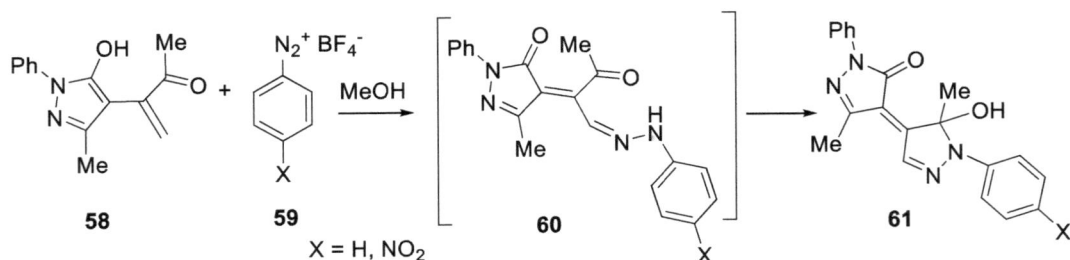

Scheme (15). Synthesis of the 3,4'-bipyrazole derivative **61**.

Heating a mixture of ethyl acetoacetate and 4-bromopyrazolone **62** gave the pyrazole ester **63** that was converted into the 4,4'-bipyrazole-5,5'-dione **64** (X = O) upon treatment with phenylhydrazine. Treatment of compound **64** (X = O) with phosphorus pentasulfide (P_2S_5) afforded compound **64** (X = S). Vilsmeier reaction of 4,4'-bipyrazole **64** (X = S, O), using DMF/POCl$_3$ reagent at 5-10 °C, produced the 5-chloro-4,4'-bipyrazole derivatives **65**. However, conducting the same reaction at 70-75 °C led to the formation of the fused-tricyclic thieno[2,3-c:5,4-c']dipyrazole derivative **66** (Scheme **16**) [38 - 40].

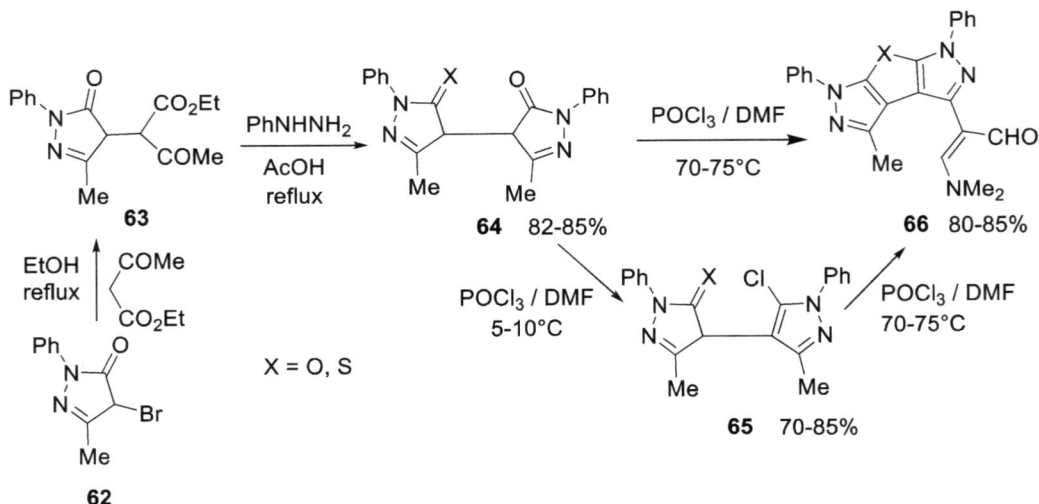

Scheme (16). Synthesis of the 3,4'-bipyrazole derivative **66**.

Selective electrophilic nitration of the unsubstituted 4,4`-bipyrazole **4** was carried out in dilute HNO_3 and afforded the mononitrated product **67** in a very high yield. The mononitrated structure **68** showed five singlet signals in its ^1H NMR spectrum at δ 7.85, 8.09, 8.26 (for bipyrazole-CH protons) and at 13.00 and 13.98 (for bipyrazole-NH protons); and its ^{13}C NMR had six C-peaks at δ 151.7, 138.6, 130.3, 127.9, 109.9, 109.1. The symmetric 3,3`-dinitro-4,4`-bipyrazole (**68**) was obtained in a high yield when **4** was heated with nitric acid at 140 °C in the presence of 80% phosphoric acid. *C*-Nitration for the generation of 3,5-dinitro-(**69**) was isolated in 100% yield using 2.2 eq. HNO_3, 91% H_2SO_4 at 170 °C for 48 h, however, the 3,3`,5,5`-tetranitro-4,4`-bipyrazole **71** was obtained, in 100% yield, using excess nitric acid and 91% $H_2SO_4$ at 170 °C for 48 h. Bipyrazole **4** was also turned into the trinitro-4,4`-bipyrazole **70** (in 94% yield) by nitration using HNO_3 and H_3PO_4 (Scheme **17**) [8, 9]. The dinitrated structure **68** showed only two singlet signals in its ^1H NMR spectrum at δ 14.13 (NH) and 8.20 (CH) and its ^{13}C NMR showed three signals at δ 153.0, 132.5, 106.9; however, the ^1H NMR of the dinitrated structure **69** showed only one singlet signal at δ 7.93 and its ^{13}C NMR had four C-peaks at δ 148.2, 135.6 107.2, 104.5

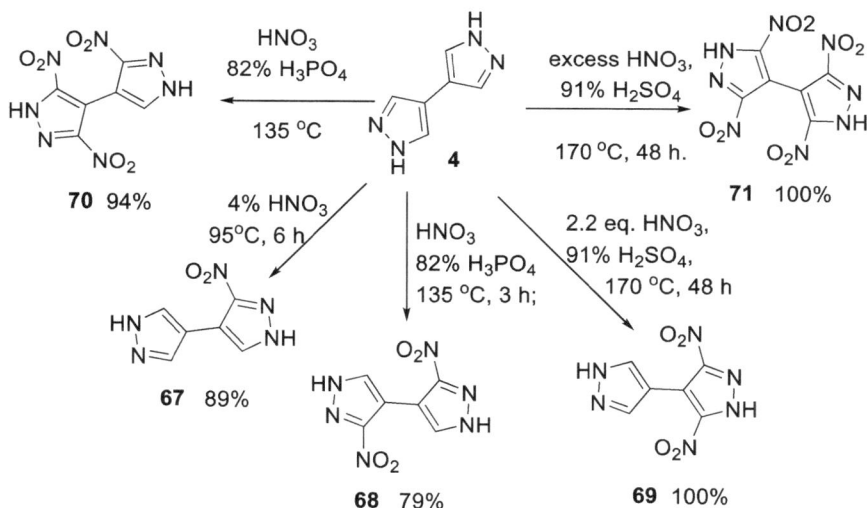

Scheme (17). Selective nitro functionalization of the 4,4`-bipyrazoles **67-71**.

CONCLUSION

In this chapter, all synthetic approaches for the construction of 4,4`-bipyrazole systems were explored. Among the reported synthetic routes are cyclocondensation of dicarbonyl-based 4-pyrazoles or tetraketones or bis-enals with hydrazines as well as the 1,3-dipolar cycloadditions in addition to homocoupling dimerization reactions. The reported 4,4`-bipyrazole derivatives possessed high biological potencies and industrial applications. Some of their metal complexes, such palladium(II) and platinum(II) had potential anticancer activities, and their gadolonium(III) complexes showed clinical Magnetic Resonance Imaging (MRI) properties. More in-depth application study is recommended to explore the interesting veiled applications of 4,4`-bipyrazoles.

REFERENCES

[1]   Assad, A. Treatment of hematological malignancies using combination of a Janus kinase 1 (JAK1) inhibitor, an immunomodulatory agent and a steroid, U.S. *Pat. Appl. Publ,* US 20200129517 A1.

[2]   Li, Y.L; Zhuo, J.; Qian, D.Q.; Mei, S.; Cao, G.; Pan, Y.; Li, Q.; Jia, Z. Preparation of bipyrazole derivatives as JAK kinase inhibitors. *PCT Int. Appl.,* WO 2014186706 A1.

[3]   Igarashi, T.; Sakurai, K.; Oi, T.; Obara, H.; Ohya, H.; Kamada, H. New sensitive agents for detecting singlet oxygen by electron spin resonance spectroscopy. *Free Radic. Biol. Med.,* **1999**, *26*(9-10), 1339-1345.
[http://dx.doi.org/10.1016/S0891-5849(98)00291-3] [PMID: 10381208]

[4]   Ohara, H.; Igarashi, T.; Sakurai, K.; Oshii, T. Bipyrazole derivative, and medicine or reagent comprising the same as active component. US 6121305A (2000).

[5]   Ohara, H.; Igarashi, T.; Sakurai, K.; Oshii, T. Bipyrazole derivative, and medicine and reagent consisting essentially thereof, JP 10306077A (1998).

[6]   Saha, N.; Misra, A. Design, synthesis and spectroscopic characterisation of palladium(II) and platinum(II) complexes of 5,5′-dimethyl-3,3′-bipyrazole with potential anti-tumor properties. *J. Inorg. Biochem.,* **1995**, *59*(2-3), 234.
[http://dx.doi.org/10.1016/0162-0134(95)97340-V]

[7]   Mayoral, E.P.; García-Amo, M.; López, P.; Soriano, E.; Cerdán, S.; Ballesteros, P. A novel series of complexones with bis- or biazole structure as mixed ligands of paramagnetic contrast agents for MRI. *Bioorg. Med. Chem.,* **2003**, *11*(24), 5555-5567.
[http://dx.doi.org/10.1016/j.bmc.2003.07.002] [PMID: 14642600]

[8]   Gospodinov, I.; Domasevitch, K.V.; Unger, C.C.; Klapötke, T.M.; Stierstorfer, J. Stierstorfer, Midway between Energetic Molecular Crystals and High-Density Energetic Salts: Crystal Engineering with Hydrogen Bonded Chains of Polynitro Bipyrazoles. *J. Cryst. Growth Des.,* **2020**, *20*(2), 755-764.
[http://dx.doi.org/10.1021/acs.cgd.9b01177]

[9]   Domasevitch, K.V.; Gospodinov, I.; Krautscheid, H.; Klapötke, T.M.; Stierstorfer, J. Facile and selective polynitrations at the 4-pyrazolyl dual backbone: straightforward access to a series of high-density energetic materials. *New J. Chem.,* **2019**, *43*(3), 1305-1312.
[http://dx.doi.org/10.1039/C8NJ05266B]

[10]   Pettinari, C.; Tăbăcaru, A.; Galli, S. Coordination polymers and metal–organic frameworks based on poly (pyrazole)-containing ligands. *Coord. Chem. Rev.,* **2016**, *307*, 1-31.
[http://dx.doi.org/10.1016/j.ccr.2015.08.005]

[11]   Taylor, M.K.; Juhl, M.; Hadaf, G.B.; Hwang, D.; Velasquez, E.; Oktawiec, J.; Lefton, J.B.; Runčevski, T.; Long, J.R.; Lee, J.W. Palladium-catalyzed oxidative homocoupling of pyrazole boronic esters to access versatile bipyrazoles and the flexible metal-organic framework Co(4,4′-bipyrazolate). *Chem. Commun. (Camb.),* **2020**, *56*(8), 1195-1198.
[http://dx.doi.org/10.1039/C9CC08614E] [PMID: 31898719]

[12]   Ponomarova, V.V.; Komarchuk, V.V.; Boldog, I.; Chernega, A.N.; Sieler, J.; Domasevitch, K.V. Mixed-anion complexes with a bipyrazolyl ligand. A new entry to a realm of three-dimensional five-connected coordination topologies. *Chem. Commun. (Camb.),* **2002**, (5), 436-437.
[http://dx.doi.org/10.1039/b110599j] [PMID: 12120529]

[13]   Boldog, I.; Rusanov, E.B.; Sieler, J.; Blaurock, S.; Domasevitch, K.V. Construction of extended networks with a trimeric pyrazole synthon. *Chem. Commun. (Camb.),* **2003,** (6), 740-741. [http://dx.doi.org/10.1039/b212540d] [PMID: 12703800]

[14]   Boldog, I.; Sieler, J.; Domasevitch, K.V. A unique polymeric coordination system that exhibits supramolecular isomerism within two dimensions. *Inorg. Chem. Commun.,* **2003,** *6*(6), 769-772. [http://dx.doi.org/10.1016/S1387-7003(03)00101-1]

[15]   Kruger, P.E.; Moubaraki, B.; Fallon, G.D.; Murray, K.S. Tetranuclear copper(II) complexes incorporating short and long metal–metal separations: synthesis, structure and magnetism. *J. Chem. Soc., Dalton Trans.,* **2000,** (5), 713-718. [http://dx.doi.org/10.1039/a908177a]

[16]   Zhang, Z-X.; Huang, H.; Yu, S-Y. Synthesis and Structure of a Dipyrazol-bridged Macrocyclic Palladium (II) Complex. *Wuji Huaxue Xuebao,* **2004,** *20*, 849-852.

[17]   Yu, S-Y.; Huang, H-P.; Li, S-H.; Jiao, Q.; Li, Y.Z.; Wu, B.; Sei, Y.; Yamaguchi, K.; Pan, Y.J.; Ma, H.W. Solution self-assembly, spontaneous deprotonation, and crystal structures of bipyrazolate-bridged metallomacrocycles with dimetal centers. *Inorg. Chem.,* **2005,** *44*(25), 9471-9488. [http://dx.doi.org/10.1021/ic0509332] [PMID: 16323935]

[18]   Domasevitch, K.V.; Boldog, I.; Rusanov, E.B.; Hunger, J.; Blaurock, S.; Schröder, M.; Sieler, J. Helical bipyrazole networks conditioned by hydrothermal crystallization. *Z. Anorg. Allg. Chem.,* **2005,** *631*, 1095-1100. [http://dx.doi.org/10.1002/zaac.200400515]

[19]   He, J.; Yin, Y-G.; Wu, T.; Li, D.; Huang, X-C. Design and solvothermal synthesis of luminescent copper(I)-pyrazolate coordination oligomer and polymer frameworks. *Chem. Commun. (Camb.),* **2006,** (27), 2845-2847. [http://dx.doi.org/10.1039/b601009a] [PMID: 17007392]

[20]   Zhang, J-P.; Horike, S.; Kitagawa, S. A flexible porous coordination polymer functionalized by unsaturated metal clusters. *Angew. Chem. Int. Ed.,* **2007,** *46*(6), 889-892. [http://dx.doi.org/10.1002/anie.200603270] [PMID: 17183498]

[21]   Zhang, J.P.; Kitagawa, S. Supramolecular Isomerism, Framework Flexibility, Unsaturated Metal Center, and Porous Property of Ag(I)/Cu(I) 3,3',5,5'-Tetrametyl-4,4'-Bipyrazolate. *J. Am. Chem. Soc.,* **2008,** *130*(3), 907-917. [http://dx.doi.org/10.1021/ja075408b] [PMID: 18166049]

[22]   Zhang, E.; Jia, Q.; Zhang, J.; Ji, Z. Metal-Anion Coordination and Linker-Anion Hydrogen Bonding in the Construction of Metal-Organic Frameworks from Bipyrazole. *Chin. J. Chem.,* **2016,** *34*(2), 191-195. [http://dx.doi.org/10.1002/cjoc.201500640]

[23]   Sun, Y.Q.; Deng, S.; Liu, Q.; Ge, S.Z.; Chen, Y.P. A green luminescent 1-D helical tubular dipyrazol-bridged cadmium(II) complex: a coordination tube included in a supramolecular tube. *Dalton Trans.,* **2013,** *42*(29), 10503-10509. [http://dx.doi.org/10.1039/c3dt50620g] [PMID: 23752348]

[24]   Krylov, I.B.; Budnikov, A.S.; Lopat'eva, E.R.; Nikishin, G.I.; Terent'ev, A.O. Mild Nitration of Pyrazolin-5-ones by a Combination of Fe(NO₃)₃ and NaNO₂ : Discovery of a New Readily Available Class of Fungicides, 4-Nitropyrazolin-5-ones. *Chemistry,* **2019,** *25*(23), 5922-5933. [http://dx.doi.org/10.1002/chem.201806172] [PMID: 30834586]

[25]   Bruno, G.; Grassi, G.; Nicolo, F.; Risitano, F.; Scopelliti, R. 4-[5'-Oxo-1',3'-diphenyl-4'-(p-tolylmethyl)]-1,3-diphenyl-4-(p-tolylmethyl)pyrazol-5-one: Product of a Reductive Dimerization Reaction. *Acta Crystallogr.,* **1996,** *C52*, 3129-3131.

[26]   Risitano, F.; Grassi, G.; Caruso, F.; Foti, F. *C,C*- and *C,N*-linked dimers and 4-arylmethyl derivatives from 4-arylmethylene pyrazol-5-ones and isoxazol-5-ones with 2-arylbenzimidazolines. *Tetrahedron,*

1996, *52*(4), 1443-1450.
[http://dx.doi.org/10.1016/0040-4020(95)00970-1]

[27]   Eller, G.A.; Vilkauskaitė, G.; Šačkus, A.; Martynaitis, V.; Mamuye, A.D.; Pace, V.; Holzer, W. An unusual thionyl chloride-promoted C-C bond formation to obtain 4,4′-bipyrazolones. *Beilstein J. Org. Chem.,* **2018**, *14*(1), 1287-1292.
[http://dx.doi.org/10.3762/bjoc.14.110] [PMID: 29977396]

[28]   Taisei, V.; Hideshi, M.; Keiko, N.; Nariichi, O.; Isoo, I. *Nagoya Shicitsu Daigaku Yakyza Kubu Kenkye Nemro,*1981, *29*, 25. *Chem. Abstr.,* **1981**, *98*, 143319.

[29]   Zeng, C.; Zhang, Z.; Zhong, R.; Hu, L. Method for electrochemically preparing 5,5'-dihydroxyl-4,4'-dipyrazole compound. *Faming Zh. Shenq,* **2010**. CN 101838816, A.

[30]   Sharko, A.V.; Senchyk, G.A.; Rusanov, E.B.; Domasevitch, K.V. Preparative synthesis of 3 (5), 3′(5′)-dimethyl-4, 4′-bipyrazole. *Tetrahedron Lett.,* **2015**, *56*(44), 6089-6092.
[http://dx.doi.org/10.1016/j.tetlet.2015.09.072]

[31]   Ibrahim, M.K.A.; El-Reedy, A.M.; El-Gharib, M.S.; Farag, A.M. Activated Nitrile in Heterocyclic Synthesis. A Novel Synthesis of Pyrazolo[3,4-b]pyridine, Pyrrolo[2,3-c]pyrazole, Pyrano[2,3-c]pyrazole, and Pyrrolo[3,4-c]pyrazole. *J. Indian Chem. Soc.,* **1987**, *64*, 345-347.

[32]   Jin, S.; Huang, Y.; Fang, H.; Wang, T.; Ding, L. 3,5,3′,5′-Tetramethyl-4,4′-bi(1H-pyrazolyl) hemihydrate. *Acta Crystallogr.,* **2012**, *E68*(10), o2896-o2896.

[33]   Rey, M.; Beaumont, S. Molybdenum-Mediated One-Pot Synthesis of Pyrazoles from Isoxazoles. *Synthesis,* **2019**, *51*(20), 3796-3804.
[http://dx.doi.org/10.1055/s-0039-1690615]

[34]   Attanasi, O.A.; Filippone, P.; Fiorucci, C.; Mantellini, F. Reaction between Conjugated Azoalkenes and Pyrazolinones: A Precious Entry to New Conjugated Azodiene and Asymmetric 4, 4′-Bipyrazole Derivatives. *Chem. Lett.,* **2020**, *29*(9), 984-985.
[http://dx.doi.org/10.1246/cl.2000.984]

[35]   Bakavoli, M.; Feizyzadeh, B.; Rahimizadeh, M. Investigation of hydrazine addition to functionalized furans: synthesis of new functionalized 4,4′-bipyrazole derivatives. *Tetrahedron Lett.,* **2016**, *47*(50), 8965-8968.
[http://dx.doi.org/10.1016/j.tetlet.2006.10.037]

[36]   El-Gendy, A.M.; Deeb, A.; El-Safty, M.; Said, S.A. *Egypt. J. Chem.,* **1989**, *32*, 335.

[37]   Lyčka, A.; Mustroph, H. ^{1}H, ^{13}C and ^{15}N NMR spectra of the reaction product of benzenediazonium fluoroborates with 1-phenyl-3-methyl-4-(α-acet-ethylidene)-pyrazol-5-one. *Dyes Pigments,* **1997**, *34*(2), 101-107.
[http://dx.doi.org/10.1016/S0143-7208(96)00069-1]

[38]   Das, N.B.; Mittra, A.S. Heterocyclic fungicides. Part III. *J. Indian Chem. Soc.,* **1978**, *55*, 829-831.

[39]   Awad, I.M.A. Studies in Vilsmeier–Haack Reaction. IX. Synthesis and Application of Novel Heterocyclo-Substituted Furo [2,3-c: 5,4-c′] dipyrazole Derivatives. *Bull. Chem. Soc. Jpn.,* **1992**, *65*(6), 1652-1656.
[http://dx.doi.org/10.1246/bcsj.65.1652]

[40]   Awad, I.M. Studies on the vilsmeier-haack reaction. part xii: novel heterocyclo-substituted thieno [2,3-c:5,4-c′] dipyrazole derivatives. *Phosphorus Sulfur Silicon Relat. Elem.,* **1992**, *72*(1-4), 81-91.
[http://dx.doi.org/10.1080/10426509208031542]

CHAPTER 5

Applications of Bipyrazole Derivatives

Abstract: Numerous bipyrazole-based metal-organic frameworks (MOF) were synthesized *via* mixing a number of bipyrazole ligands with several transition-metal cations, and the obtained MOF represented interesting applications in the field of material science and pharmaceuticals due to their high degree of crystallinity and internal porosity. There are photo-luminescence, sensing, gas separations, electrical conductivity, and energy storage, among those interesting applications.

Keywords: Bipyrazoles, Energetic organic materials, Gas separation, MOF, Nitropyrazoles, OLED.

1. INTRODUCTION

Recently, bipyrazole-based *metal coordination compounds* displayed interesting applications in pharmaceuticals and in material science. For example, gold(III) and ruthenium(II) complexes of bipyrazoles were proved to be anticancer agents [1, 2], and copper(I) complexes had excellent antibacterial activity [3], whereas gold(III), platinum(II), osmium(II) and copper(I) complexes were involved in the fabrication of luminescence Organic Light-Emitting Diodes (OLED) and laser materials [4 - 7]. Bipyrazole ligands coordinate up to four different metal centers to give three-dimensional structures known as metal–organic frameworks (MOFs). Such MOFs had promising wide applications in drug delivery, sensing, gas separations, electrical conductivity, and energy storage due to their high degree of crystallinity and internal porosity [8, 9]. Bipyrazoles (especially Bippyphos) played an important role as ligands for palladium-catalyzed cross-coupling reactions of aryl halides [10 - 13].

2. APPLICATIONS OF BIPYRAZOLE DERIVATIVES

2.1. Bipyrazoles as Ligands

The 3,3`-bipyrazole-based Pd(II)-complex **1** was synthesized and reported as an efficient precatalyst for Suzuki-Miyaura C-C cross-coupling reactions of aryl halides with arylboronic acids in aqueous media [14].

Kamal M. Dawood and Ashraf A. Abbas

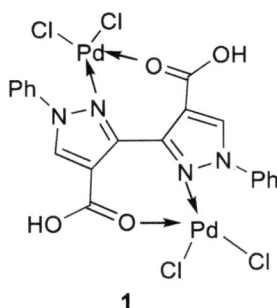

1

5-(Di-*tert*-butylphosphino)-1′,3′,5′-triphenyl-1′*H*-[1,4′]bipyrazole (Bippyphos) (**2**) was reported as an efficient co-catalyst in the palladium-catalyzed hydroxylation of several (hetero)aryl halides **3** under mild conditions as well as in the synthesis of substituted benzofurans and related heteroaromatic derivatives [11] (Scheme **1**).

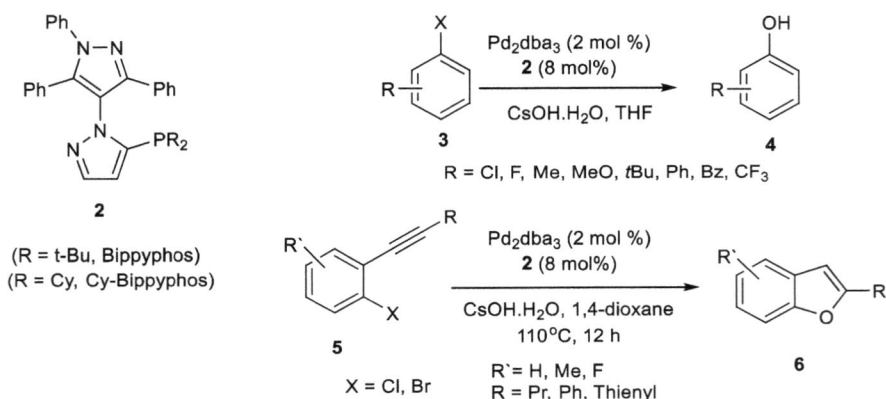

Scheme (1). Synthesis of hydroxyl compounds **4** and substituted benzofurans **6**.

The bipyrazole derivatives (bippyphos) **2** were applied as efficient ligands in the palladium-catalyzed C-O and C-N cross-coupling reactions of aryl halides with primary alcohols and with urea derivatives, respectively [12, 13, 15 - 17].

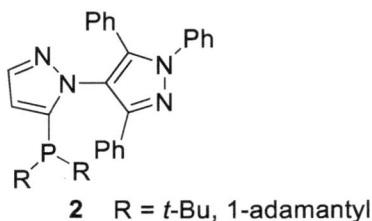

2   R = *t*-Bu, 1-adamantyl

Polycondensation of 5,5'-dimethyl-3-chloromethyl-1,3'-bipyrazole **7** was achieved in refluxing benzene in the presence of 50% NaOH solution and led to the formation of the polypyrazolic macrocycle **8** in 75% yield (Scheme **2**). The polypyrazolic macrocycles showed excellent complexing properties as ligands with the alkali metal cations [18].

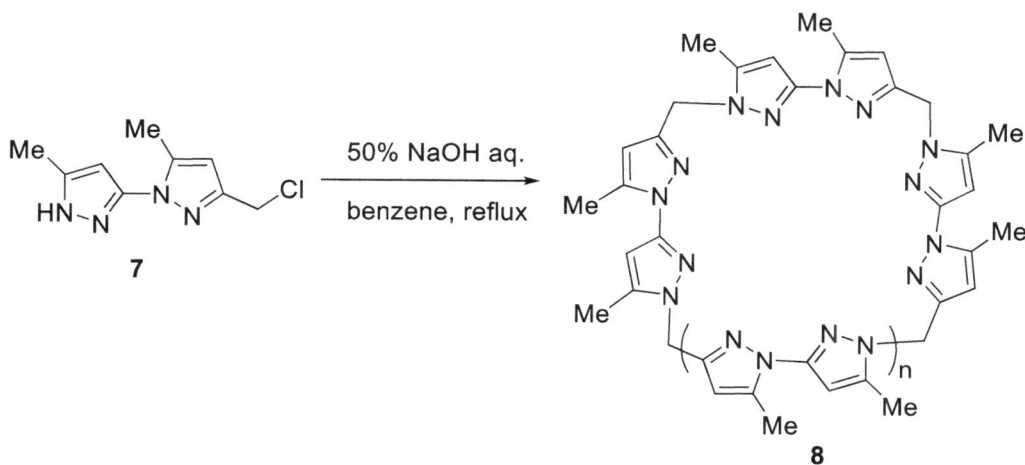

Scheme (2). Synthesis of the polypyrazolic macrocycle **8**.

The immobilized bipyrazole **9** on the surface of epoxy-silica presented good thermal stability based on the thermogravimetric analysis, and it had good binding and adsorption abilities for Hg^{2+}, Cd^{2+}, Pb^{2+}, Zn^{2+}, K^+, Na^+ and Li^+ cations [19].

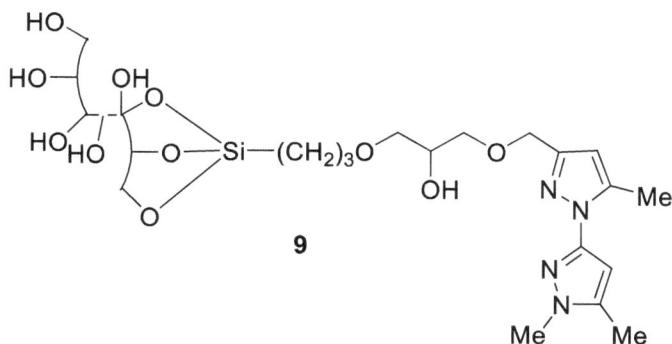

2.2. Bipyrazoles in Synthesis of Polybipyrazoles

Dehalogenative polycondensation of 3,3'-dichloro-5,5'-bipyrazoles **10** using a mixture of Ni(cod)$_2$ and 2,2'-bipyridine in DMF at 60°C resulted in the formation of poly(5,5'-bipyrazole-3,3'-diyl) derivatives **11** (Scheme **3**). The obtained polymers were characterized by their high thermal stability and electrochemical activity. The polymers **11** showed photoluminescent peaks at the onset position of the respective π–π* absorption bands [20].

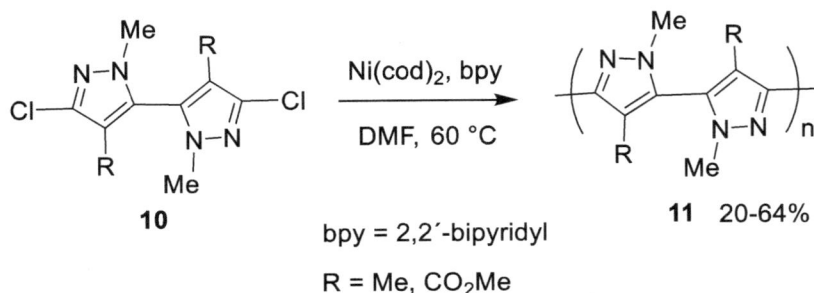

bpy = 2,2′-bipyridyl

R = Me, CO$_2$Me

Scheme (3). Synthesis of the poly(5,5′-bipyrazole-3,3′-diyl) derivatives **11**.

2.3. Bipyrazoles as Energetic Materials

Compounds **12** and **13** were reported to have potentials as high-temperature energetic materials where they had excellent thermal stabilities with decomposition temperatures; (T_d) 228 °C and 315 °C, respectively, compared with 2,4,6-triamino-1,3,5-trinitrobenzene (TATB) (T_d: 350 °C). Both compounds **12** and **13** exhibited good detonation performances, including detonation velocities and detonation pressures and low sensitivity to impact and friction thus both were classified as insensitive explosives. Compound **13** showed much higher performances than **12** due to the presence of the fused ring [21, 22].

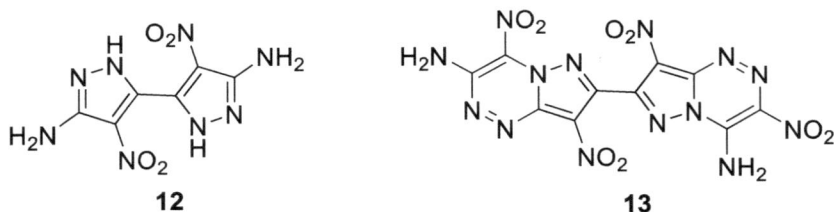

The nitrated bipyrazole derivatives **14** and **15** were reported to be metal-free primary explosives where they had higher decomposition temperatures; T_d = 376 and 365 °C, respectively, and exhibited excellent thermal stability better than hexanitrostilbene (HNS), the well-known heat resistant explosive (T_d = 320 °C). Compounds **14** and **15** had also higher densities (1.81 and 1.83 g/cm^3, respectively) better than HNS (1.745 g/cm^3). Compounds **14** and **15** were also reported as insensitive heat resistant explosives and were highly sensitive primary explosives where their impact sensitivities (IS) and friction sensitivities (FS) were; IS: > 40 J, FS: > 360 N. The detonation pressures (P) and the detonation velocities (Dv) of the heat resistant explosives **14** and **15** were (26.2 GPa, 8026 m/s) and (26.9 GPa, 8120 m/s), respectively, and had better detonation properties than HNS (P: 24.5 GPa, Dv: 7629 m/s) [23].

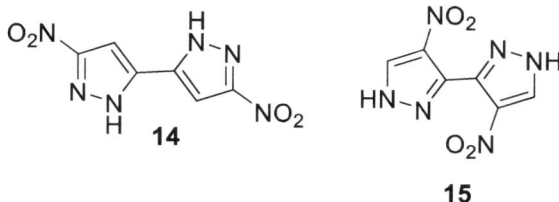

14   **15**

The energetic properties of the polynitro-bipyrazole derivative **16** and its fused system **17** showed that they had high thermal stabilities where their decomposition temperatures were 252 °C and 233 °C, respectively. The high thermal stabilities of compounds **16** and **17** were assigned due to the conjugation of double bonds. Both compounds **16** and **17** exhibited also high heats of formation. Their detonation properties were variable, where compound **17** had detonation performance (detonation velocity: 8504 m/s; detonation pressure: 31.0 GPa) little lower than RDX, however the detonation performance of compound **16** (detonation velocity: 9631 m/s; detonation pressure: 44.0 GPa) was superior to 1,3,5-trinitro-1,3,5-triazinane (RDX) (detonation velocity: 8795 m/s; detonation pressure: 34.9 GPa) [24].

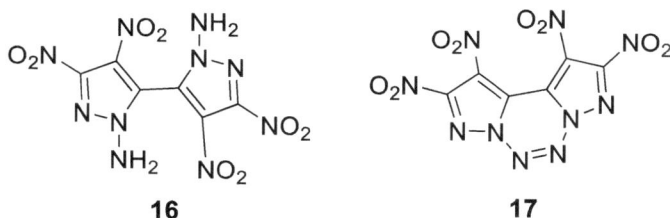

16   **17**

The polynitro-bipyrazole derivatives **18**, **19** and **20** had good thermal stabilities and decomposed at 150 °C, 228 °C and 323 °C, respectively. Thus, compound **20** had higher thermal stability than that of lead azide (315 °C), the commonly used primary explosive. The densities of the derivatives **18**, **19** and **20** were 1.882, 1.916 and 2.029 g/cm^3, respectively, and their detonation velocities were 8987, 8035 and 7769 m/s, respectively, compared to that of the well-known explosive RDX (8795 m/s). In addition, the dipotassium salt **20** was classified as a sensitive compound where its impact and friction sensitivities were 4 J and 40 N, respectively, and compounds **18** and **20** were also sensitive with impact sensitivities 5 and 6 J, respectively. Therefore, compounds **18**, **19,** and **20** were classified as high energetic density materials (HEDM). Particularly, the dipotassium salt **20** might have application in the field of primary explosives due to its high density and excellent thermal stability [25].

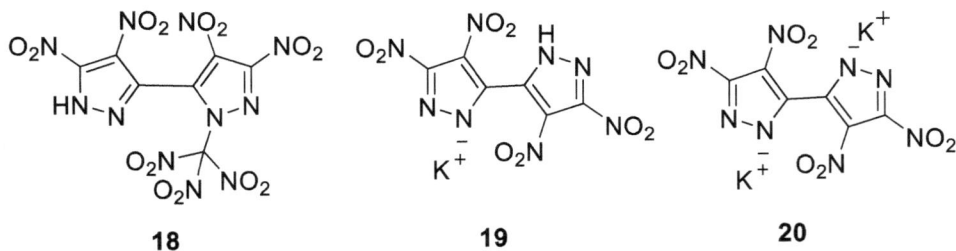

18   **19**   **20**

Various salts of the polynitro-4,4`-bipyrazoles **21** and **22** were also classified as energetic materials, where their energetic behaviors, such as detonation pressure, velocity, and energy, were determined. The monohydrate salts **21** and **22** showed high thermal stabilities with decomposition temperatures $T_{dec.}$ = 333 °C and 286 °C for the potassium **21**·H_2O and the ammonium **22**·H_2O salts, respectively, and their detonation velocities were 1.84 and 1.68 g/cm³, respectively. The energetic salt **21**·H_2O exhibited moderate impact sensitivity (7 J), however, the ammonium salt **22**·H_2O was the most insensitive material, either toward impact (40 J) or friction (>360 N) [26].

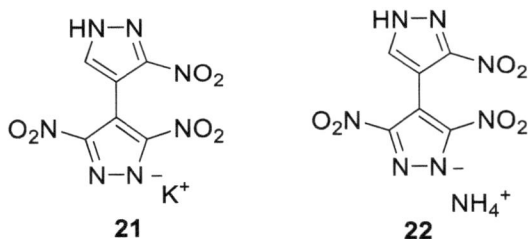

21   **22**

The physico-chemical study for the polynitrated 4,4`-bipyrazoles **23** and **24** showed that their decomposition temperatures were 314 °C and 298 °C, respectively, where increasing the number of NO_2 groups led to a little decrease of the decomposition temperatures. The density of compounds **23** and **24**, at room temperature, were 1.855 and 1.820 g/cm³, respectively. Compounds **23** and **24** exhibited positive enthalpies of formation 225 kJ/mol and 228 kJ/mol, respectively. The detonation pressures and velocities for compounds **23** and **24** were (28.6 GPa, and 8256 m/s) and (31.1 GPa, and 8520 m/s), respectively, and exceeded the reported values for 2,2',4,4',6,6'-hexanitrostilbene (HNS) (24.3 GPa, and 7612 m/s). The explosive behaviors of the trinitro- and tetranitro-bipyrazoles **23** and **24**, on a small scale, showed excellent thermal stabilities and good sensitivities and were classified as explosive materials [27].

23   **24**

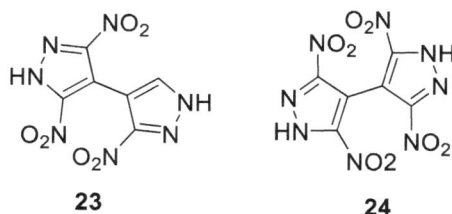

The oxygen-rich high energetic material, 4,4`,5,5`-tetranitro-2,2`-bis (trinitromethyl)-2H,20H-3,3`-bipyrazole (**25**), having ten nitro groups, was synthesized and reported to be useful as green energetic nitrogen heterocyclic material, where it has a high density (2.021 g cm^{-3} at room temperature), good positive oxygen balance, high sensitivity and high power. Thus, compound **25** might have promising applications as a high energy-density oxygen-carrier material for use in various scientific, space and military projects [28].

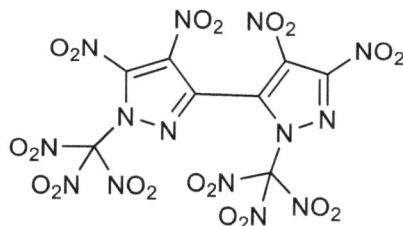

25

2.4. Bipyrazoles as Corrosion Inhibitors

1',3,5,5'-Tetramethyl-1,3'-bipyrazole **26** and 5,5'-disubstituted-3,3'-bipyrazoles **27** were used as inhibitors for the corrosion of steel in $1M$ HCl, where the inhibition efficiencies increased with increasing the inhibitors concentrations. The inhibition effect of the bipyrazoles **26** and **27** was attributed to their adsorptions at the metal–solution interface, owing to the presence of many active centers (several nitrogen atoms and many π-electrons of the pyrazole rings) for adsorption. The results revealed that inhibitive actions of bipyrazole compounds were mainly due to adsorption on steel surfaces [29 - 31].

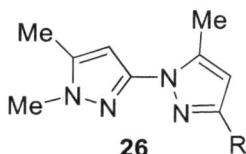

26   **27**

R = Me, CH$_2$OH, CO$_2$Et   R = Et, Ph, 4-ClC$_6$H$_4$

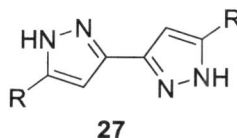

2.5. Bipyrazoles as Therapeutics

The biological and pharmacological importance of bipyrazole derivatives were extensively reported where they demonstrated a wide range of inhibitory activities against various diseases such as anti-inflammatory, anticancer, antimalarial, antimicrobial, antioxidant, antitubercular, insecticidal activities as well as enzymatic inhibitions [32].

5,5'-Dihydroxy-4,4'-bipyrazoles **28** were reported as useful medicines for the treatment of cerebral ischemia, heart diseases, gastrointestinal diseases, cancer, aging and inflammation. These medicines are useful for effectively capturing active oxygen and free radicals, which cause adult diseases where singlet oxygen generated in a photo-excited hematoporphyrin system was reacted with 5,5'-dihydroxy-4,4'-bipyrazoles **28** to give ESR signal indicating production of stable free radical [33 - 35].

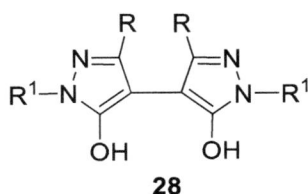

28

R = H, Me, Et, Pr, Bu, Ph

R^1 = H, Me, Et, Pr, Bu, CH_2OH, $(CH_2)_2OH$, $(CH_2)_3OH$, Ph, benzyl, naphthyl

The tetra-sodium salt of 4,4'-bipyrazole derivative **29** was synthesized and used in Gadolonium(III) complex as Paramagnetic Contrast Agent for clinical Magnetic Resonance Imaging (MRI) [36].

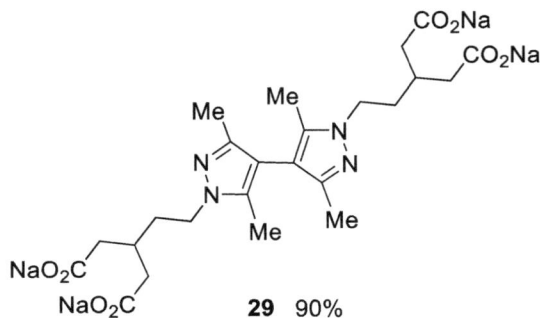

29 90%

The solvatochromic behaviors of 3,5'-bipyrazole derivatives **30** were reported in various solvents of different polarity. Spectroscopic studies showed that the solvatochromic behavior was greatly dependent on both the polarity of the medium and the hydrogen-bonding properties of the solvent. The photophysical study of 3,5'-bipyrazole derivatives **30** in different solvents helped in assessing their potential applications in different environments [37].

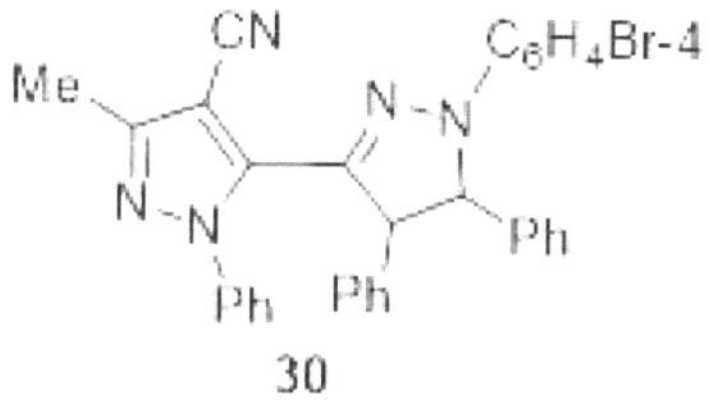

30

2.6. Bipyrazoles in Metal–organic Frameworks (MOFs)

3,3',5,5'-Tetramethyl-4,4'-bipyrazole **31** was reported as an important class of bi-heterocyclic systems accounting for its capability to form the porous coordination polymers **32** with potential uses as solid sorbents, ion exchangers, and heterogeneous catalysts [38 - 45].

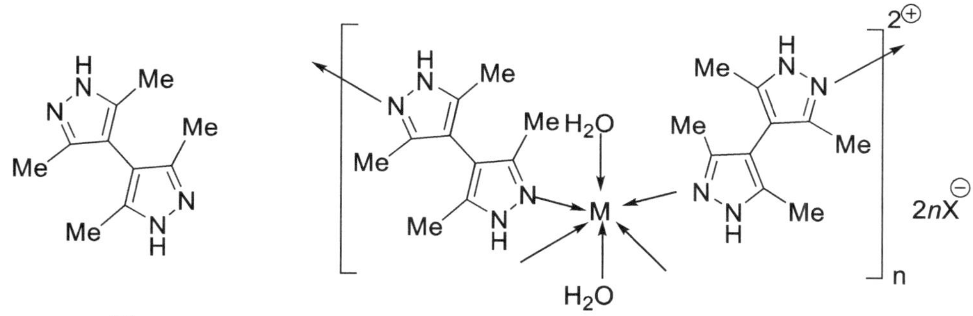

31

32   M = Cu, Co, Cd, Ni, Pd, Ag, W

Furthermore, 3,3',5,5'-tetramethyl-4,4'-bipyrazole **31** was well-studied as a hydrogen-bonding synthon as well as a neutral bidentate ligand for the synthesis of a flexible porous coordination polymer with two-coordinate Ag centers **33** (Scheme **4**) [46 - 50].

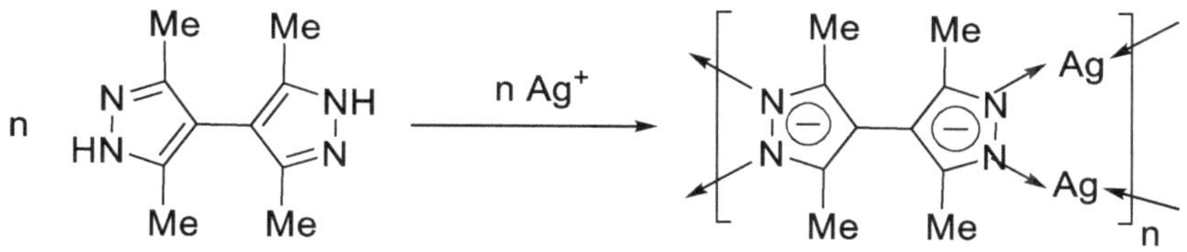

31   33

Scheme (4). Synthesis of porous coordination polymer **33**.

The carbonyl(hydrido)*bis*-(triphenylphosphane)ruthenium(II) complexes **34** were synthesized and showed good catalytic activity and transfer of hydrogen in catalyzed hydrogenation reactions [51].

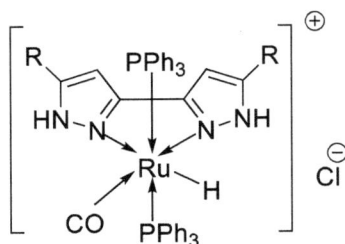

34   R = H, ^{n}Bu

3,3`,5,5`-Tetramethyl-4,4`-bipyrazole (TMBP) **31** was involved in the synthesis of three isostructural, ultramicroporous diamondoid metal–organic frameworks (MOFs), [Cu(TMBP)X]; (X=Cl, Br, I), as new benchmark C_2H_2/CO_2 separation selectivity at ambient temperature and pressure. This selectivity was attributed to the strong binding site for C_2H_2 *via* halogen$\cdots$HC interactions coupled with other noncovalent in a tight binding site of C_2H_2*versus* CO_2 [52].

A 3D Cu(I)-based metal-organic framework employing 3,3',5,5'-tetramethyl-4,4'-bipyrazolyl (H_2L) **31** and cyano ligands were found to be useful as fluorescent sensors for the detection of nitroaromatic compounds, particularly 2,4,6-trinitrophenol (TNP) [53].

Copper(I/II) complexes of 5,5'-diphenyl-1H,1'H-3,3'-bipyrazole **35** were reported to catalyze the oxidation of catechol (**36**) into o-quinone (**37**) at ambient conditions using the atmospheric oxygen as an oxidant [54] (Scheme **5**).

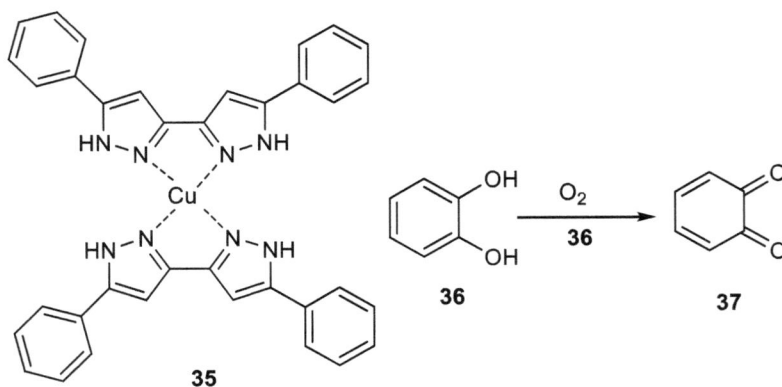

Scheme (5). Synthesis of o-quinone **37**.

The *in situ* Cu(II) complexes of three bipyrazole-based ligands: 1′,5,5′-trimethyl-1′H-1,3′-bipyrazol-3-ethyl carboxylate **38**, (1′,5,5′-trimethyl-1′H-1,3′-bipyra-ol-3-yl) methanol **39**, and 5,5′-diphenyl-3,3′-bipyrazole **40**, were synthesized. The complexes were found to catalyze the oxidation of catechol (**36**) into o-quinone (**37**) at ambient conditions using the oxygen of the atmosphere as an oxidant [55] (Scheme **6**).

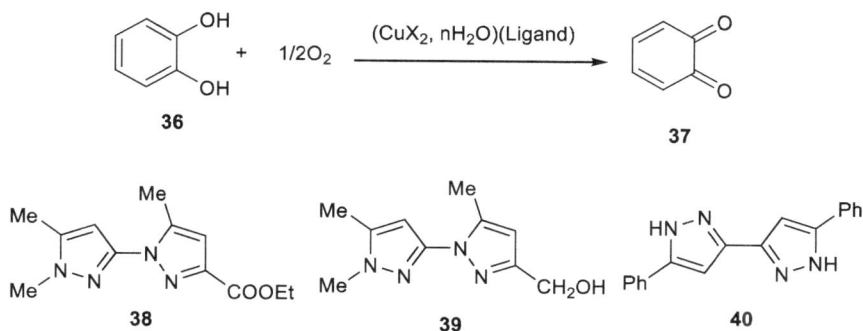

Scheme (6). Synthesis of o-quinone **37**.

The copper(I)-containing metal-organic framework; [Cu$^I_2$(phbpz)] (**42**) (H$_2$-phbpz = 3,3',5,5'-tetraphenyl-1*H*,1'*H*-4,4'-bipyrazole (**41**)) was synthesized and showed a weak binding of carbon monoxide on Cu(I) centers (Scheme **7**). The reactivity of **42** towards oxidizing agents, such as H$_2$O$_2$, *t*-BuOOH and Br$_2$ was also examined. The MOF **42** also showed luminescence upon exposure to UV radiation. The weak chemisorption of carbon monoxide on Cu(I) centers was confirmed by sorption and IR measurements however, no chemisorption of oxygen was observed. The reactions of **42** with H$_2$O$_2$ or *t*-BuOOH showed that the MOF is stable during repeated oxidation/reduction processes [56 - 58].

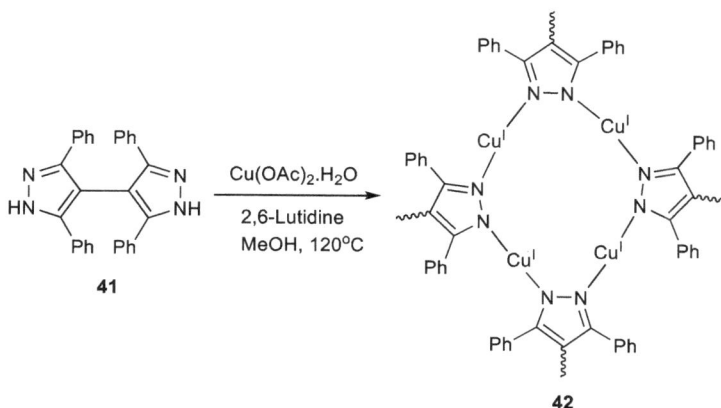

Scheme (7). Synthesis of copper(I)-containing MOF **42**.

Treatment of 3-amino-4,4'-bipyrazole (H_2BPZNH_2) (**43**) with metal acetates $M(OAc)_2 \cdot nH_2O$ afforded the metal–organic frameworks (MOFs) $M(BPZNH_2)$ (M = Zn, Ni, Cu) under solvothermal condition. The Zn(II) polymer was characterized by a 3D porous network featuring tetrahedral metallic nodes and bridging $BPZNH_2^{2-}$ anions and exhibited a high capacity of CO_2 uptake [59].

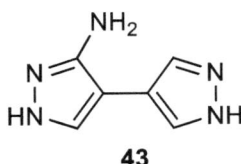

43

The pillared-layer Co(II)-based metal–organic frameworks (MOF): $[Co_2(L)(TMBP)(H_2O)]n$ (**44**) and $[Co_2(bpdc)_2(H_2TMBP)] \cdot 2(DMF) \cdot 5(H_2O)$ (**45**) was synthesized from the solvothermal reaction of 5,5'-(1,4-phenylenebis (methyleneoxy))diisophthalic acid (H_4L) (**46**) or 4,4'-biphenyldicarboxylic acid (H_2bpdc) (**47**) with cobalt(II) salts in the presence of 3,3`,5,5`-tetramethyl-4,4`-bipyrazole (TMBP) (**31**) ligand. TMPBP bridges the dicobalt-tetracarboxylate $[Co_2(O_2CR)_4]$ clusters with an angular coordination configuration forming left- and right-handed helical chains. The Co(II)-based MOF (**44**) showed efficient photocatalytic performance in the degradation of methyl violet (MV) under UV irradiation. However, the Co(II)-based MOF (**45**) showed strong affinity for CO_2 molecules and high adsorption selectivities for CO_2 over N_2 and H_2 and also exhibited strong antiferromagnetic exchange interactions between the Co^{2+} ions dinuclear clusters. The strong affinity of the framework to CO_2 interactions is attributed to the uncoordinated –NH groups of the bipyrazole system in pores through N–H---O=C=O hydrogen bonds. Moreover, the strong antiferromagnetic coupling between the intra-cluster Co^{2+} centers is accounted for by the short Co----Co distances in the cluster [60, 61].

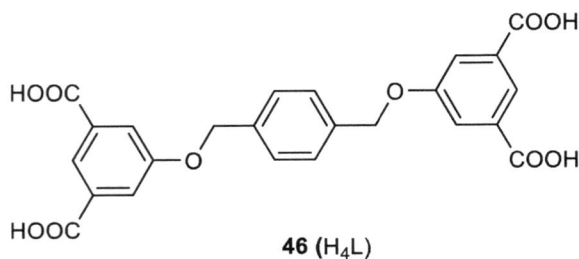

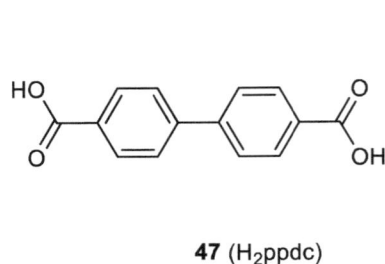

46 (H_4L)   **47 (H_2ppdc)**

The reaction of $Cd(OAc)_2 \cdot 2H_2O$ with a mixture of the ligand 5-(4-carboxyphenoxy)isophthalic acid (**48**) (H_3L) and the co-ligand 3,3',5,5'-tetramethyl-4,4'-bipyrazole (**31**) (TMBP) resulted in the formation of a thermally and chemically stable mixed ligand type Cd(II) coordination polymer **49**. The obtained mixed coordination polymer **49** was utilized as a sensor for metal cations as well as a highly active photocatalyst for the decomposition of organic dyes in polluted water [62].

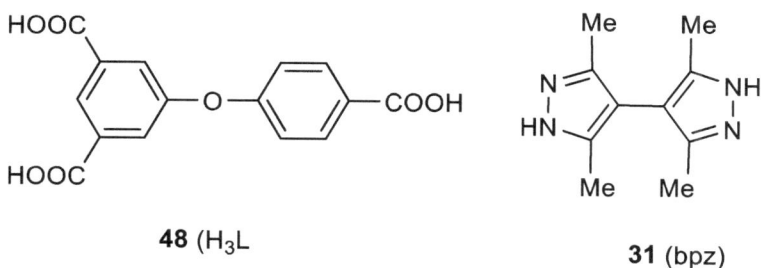

48 (H₃L

31 (bpz)

The metal–organic frameworks (MOFs); [Cd(H₂L)(TMBP)]n (**50**) and [Cd₂(H₄L)(L)(TMBP)₂]n (**51**) were obtained from 5,5`-(1,4-phenylenebis (methyleneoxy))diisophthalic acid (**52**) (H₄L) and 3,3`,5,5`-tetramethyl-4,4`-bipyrazole (**31**) (TMBP) ligand with cadmium(II) salts at two different reaction temperatures. The luminescence investigation demonstrated that both **50** and **51** had good turn-off luminescence sensing against nitroaromatic compounds (NACs), especially m-nitrophenol (MNP). The photocatalytic properties for both MOFs demonstrated that they had efficient photocatalytic performances to degrade methyl violet (MV) under UV irradiation [63].

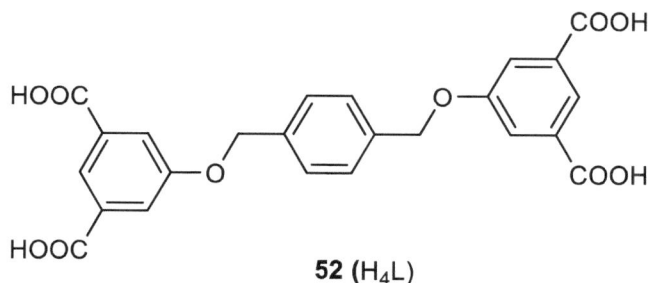

52 (H₄L)

The cadmium metal-organic framework, {[Cd(BDC)(TMBP)]·2DMF·H₂O}ₙ**53** (BDC = benzene-1,3-dicarboxylic acid (**54**)), was synthesized. The Cd-MOF **53** demonstrated a permanent porosity with high selectivity in adsorption of CO_2 gas over N_2 and CH_4 at 195 K as well as high selectivity in adsorption of water vapors over methanol and ethanol vapors at room temperature [64].

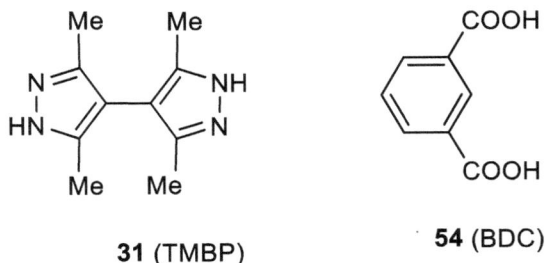

31 (TMBP)   **54 (BDC)**

Hydrothermal reaction of cadmium sulfate octahydrate with the flexible ligand, 3,3′,5,5′-tetramethyl-4,4′-bipyrazole (**31**) and the auxiliary O-donor ligand; naphthalene-4,5-dicarboxylic acid-1,8-anhydride (**56**) [obtained from *in situ* dehydration of the naphthalene-1,4,5,8-tetracarboxylic acid (**55**) in the hydrothermal reaction (Scheme **8**)] resulted in the formation of the coordination polymer, $\{[Cd_2(TMBP)(ntcaa)_2(H_2O)]\cdot H_2O\}n$ (**57**). The polymer complex **57** demonstrated a 2D wave-like layer structure based on tetranuclear Cd_4O_4 cluster units. Complex **57** also showed a luminescent property in the solid-state at room temperature. The results established that the coordination mode of the TMBP ligand **31** and the auxiliary fused-ring aromatic multicarboxylic acids played an essential role in the formation of coordination frameworks [65].

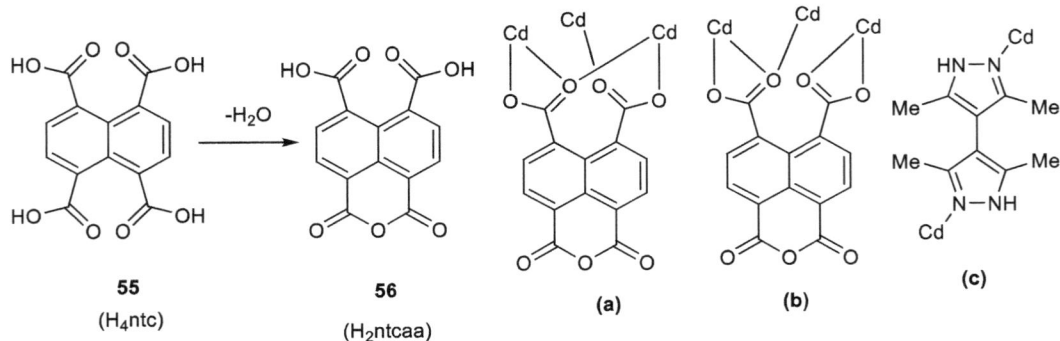

55   **56**   (a)   (b)   (c)
(H$_4$ntc)   (H$_2$ntcaa)

Scheme (8). Synthesis of **56** *via* the dehydration of **55** and the coordination modes (**a-c**) of ligands in complex **57**.

Solvothermal reaction of $Cd(NO_3)_2\cdot 4H_2O$ with TMBP and H$_3$BTC in ethanol afforded the 3D porous coordination polymer $\{[Cd(HBTC)(TMBP)]\cdot 3EtOH\}_n$ (**58**) (H$_3$BTC = benzene-1,3,5-tricarboxylic acid **59**). Complex **58** was thermally stable microporous MOF and the pores of the polymer were decorated with free carboxylic acid groups decorating the channels with potential nitrogen gas adsorption and solid state emission properties [66, 67].

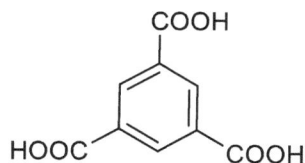

59 (H₃BTC)

The 3,3',5,5'-tetramethyl-4,4'-bipyrazole **31** (TMBP) was employed in the synthesis of two two-dimensional metal-organic frameworks (MOF) structures; [Cu(TMBP)₂(NO₃)₂] **60** and [Zn(TMBP)₂SO₄] **61**. For the latter structure, each Zn(II) was coordinated with two nitrogen atoms from two bipyrazole ligands and two oxygen atoms from two sulfate groups to form the distorted tetrahedron structure **62** [68 - 70].

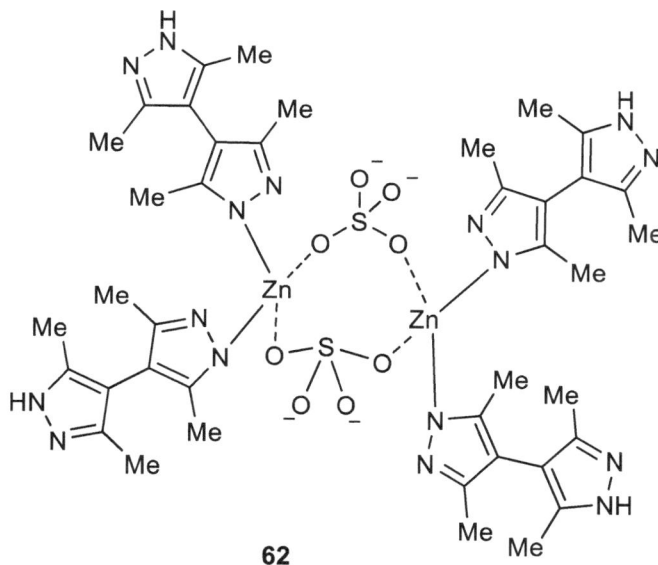

62

The zinc(II) 3D-porous MOF; [Zn₄(TMBP)₂(bpdc)₃]ₙ (**63**) employing the two rigid ligands: biphenyl-4,4'-dicarboxylic acid **47** (bpdc) and 3,3',5,5'-tetramethyl- 4,4'-bipyrazole **31** (TMBP), was synthesized. The obtained MOF had high thermal and chemical stabilities with triple aromatic walls through the combination of rigid dicarboxylate and bipyrazole [71].

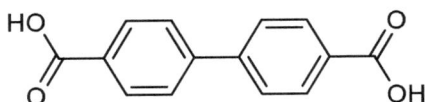

47 (bpdc)

Similarly, the Zn-based metal organic framework (MOF); $[Zn(L)0.5(TMBP)\cdot 2H_2O]_n$ (**64**), was synthesized *via* solvothermal reaction conditions. The photocatalytic study revealed that it was useful as a photocatalyst for the photodecomposition of aromatic dyes such as methyl violet (MV) under UV light [72 - 74].

A series of mixed-ligand Zn(II) MOFs containing 4,4'-bis(pyrazolate) spacers bearing different functions (unsubstituted, NO_2 or NH_2) **65-67** were prepared and found to have interesting CO_2 uptake capacity for carbon capture and sequestration (CCS) applications. The presence of NH_2 function strongly increased the amount of CO_2 adsorbed in contrast to the nitro substituent [75].

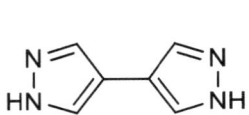 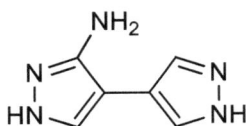 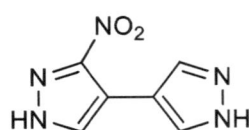

65   **66**   **67**

Two chiral coordination polymers $[Cu(I)(TMBP)(CH_3CN)_2\cdot ClO_4]_n$ (**68**) and $[Zn(II)(TMBP)(SCN)_2]n$ (**69**) (where TMBP = 3,3',5,5'-tetramethyl-4,4'-bipyrazole (**31**)) were reported. The bulk sample of compound **68** showed a negative solid-state Cotton effect, while that of **69** exhibited a positive one. However, the crystal sample of **68** showed a strong ligand-based and metal-ligand charge transition (MLCT) emission in the solid-state luminescence in contrast to the weakly luminous **69** at room temperature [76, 77].

Solvothermal synthesis of the coordination polymers [M(TMBP)] (M = Zn, Co, Cd, Cu) was reported as 3D-porous frameworks and demonstrated remarkable thermal robustness. Variable-temperature X-ray powder diffraction (VT-XRPD) measurements showed that all the four materials demonstrated: (i) permanent porosity and (ii) preserved thermal stability and framework topology along with consecutive heating−cooling series. Adsorption measurements of N_2 and CO_2 at 77 and 273 K, respectively, were studied to probe the permanent porosity of the materials and to give a coherent picture of their textural properties [78]. It was found that N_2 adsorption took preferentially place in larger micropores and the mesopores at 77 K.

The reaction of the conjugated multidentate ligand; 5,5'-(2-pyridyl)-1H,1'H-3,3'-bipyrazole **70** with different copper salts under solvothermal condition resulted in the construction of conducting materials of hexameric clusters through extended delocalization of an electron at the surface. The hexameric clusters exhibited rare stability in three valence ground states (Cu^{II}_6, $Cu^{II}_5Cu^I$, and $Cu^{II}_4Cu^I_2$). The $\pi-\pi$ orbital overlap of the ligands surrounding the hexameric copper core carriers promoted an extended electron delocalized network [79].

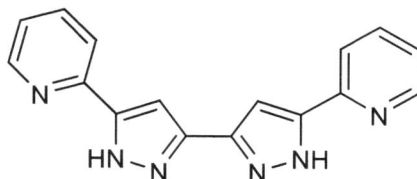

70

The metal–organic frameworks (MOFs) M(BPZNO₂) (M = Co, Cu, Zn) bearing the nitro-functionalized spacer: 3-nitro-4,4'-bipyrazole (H₂BPZNO₂) (**71**), were synthesized and showed good thermal stability with permanent porosity. The Zn(II) and Co(II) complexes demonstrated the highest CO_2 uptake [80].

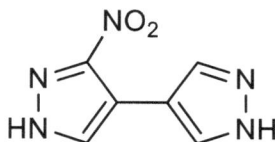

71

The coordination polymers M(Me₂BPZ) (M = Co, Zn) employing 3,3'-dimethyl-1*H*,1'*H*-4,4'-bipyrazole (H₂Me₂BPZ) (**72**) were prepared and their textural properties were examined by N_2 and CO_2 adsorption. The textural properties of the two MOFs M(Me₂BPZ) were examined by N_2 and CO_2 adsorption at 77 and 273 K, respectively, and compared to those of the non-methylated MOF [M-BPZ]. In both series, the increase of the micropore volume concomitant to the decrease of the adsorption energy passing from N_2 to CO_2 confirmed a different adsorption path for the two gases. The presence of methyl groups was found to have a positive effect in CO2 adsorption [81].

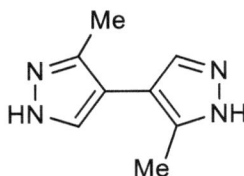

72

A 3D pillar-layered MOF [Co$_{1.5}$(BDC)$_{1.5}$(TMBP)]·DMF·4H$_2$O with **Co$_3$** clusters and two kinds of channels were synthesized using TMBP **31** and H$_2$BDC ligands (where H$_2$BDC = 1,4-benzenedicarboxylic acid (**73**)). The MOF structure exhibited moderate CO$_2$/CH$_4$ and CO$_2$/N$_2$ adsorption selectivity. Loading of the molecular iodine to the MOF greatly improved the electric conductivity of the complex to be three times compared to the MOF without molecular iodine. The molecular iodine in the MOF enhanced its conductivity by inducing n→σ* charge transfer by increasing the interactions between iodine-guest molecules and the phenyl-CH groups on the porous walls of the framework. In addition, the sorption selectivity for carbon dioxide over both methane and nitrogen was referred to two reasons: i) the ease pass of carbon dioxide in and out the pores of the framework due to its smaller kinetic diameter (3.3 Å) compared to methane (3.8 Å) and nitrogen (3.6 Å); ii) the larger quadrupole moment and the higher polarizability value of carbon dioxide compared with methane and nitrogen leading to stronger interactions between the framework and CO$_2$ molecules [82].

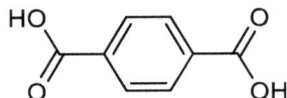

73 (H$_2$BDC)

Pt(II) complexes **74** bearing imidazolylidene–pyridylidene (impy) and 3,3`-bipyrazole ligands were synthesized and showed distinctive solid-state photophysical properties depending on the alkyl substituents. Some of the Pt-complexes showed strong solid state emissions with high suitability in serving as potential OLED phosphors, however, all Pt(II) complexes were non-emissive in solution state at room temperature [5, 83]. For example, the polycrystals of **74** (R$_1$= Me, R= C$_2$F$_5$) showed emission at 466 nm with a slight blue-shift than **74** (R$_1$= Me, R= CF$_3$) due to the more electron-withdrawing C$_2$F$_5$ groups. The emission was greatly dependent on the morphological shape of the sample, where the intensity of the peak at 466 nm, of sample **74**, sharply decreased after pulverization of the sample and the main emission was red-shifted to 569 nm.

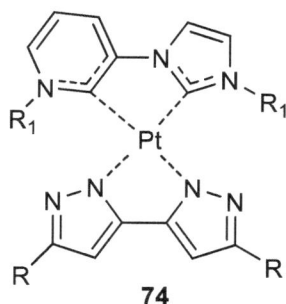

74

R = CF$_3$, C$_2$F$_5$

R$_1$ = Me, Et, iPr

Square pyramidal oxovanadium(IV) frameworks of 1`,2-diphenyl-5-(2-pyridyl)-3,4'-bipyrazole **75** were synthesized and were found to have cytotoxic activity against brine shrimp with LD$_{50}$ values obtained in the range 8–24 µg ml^{-1} [84].

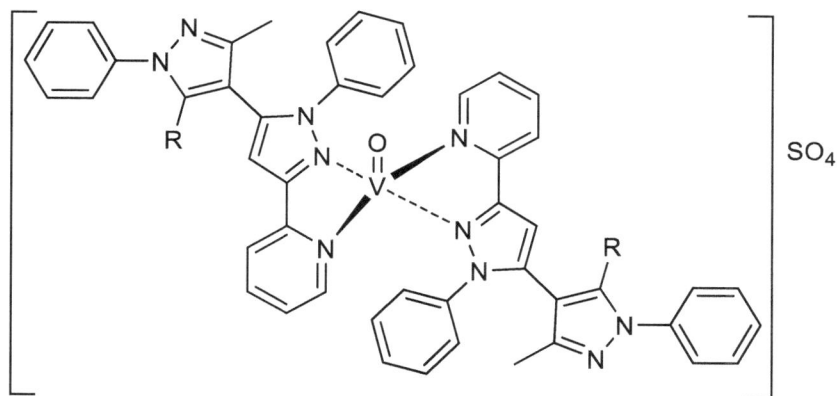

75   R = H, Cl, PhO

Iridium(III) complexes **76** with tetradentate chelates bearing tripodal arranged terpyridine and bipyrazole units were synthesized and employed in the preparation of highly efficient organic light-emitting diodes (OLEDs). This is attributed to the stronger metal–ligand bonding interaction. Ir-complexes **76** were weakly emissive in solution at room temperature and the emission efficiency highly increased upon changing to solid state. Furthermore, emission of Ir-complex **76** (R = t-Bu) occurred at slightly higher energy than that of **76** (R = H), due to the presence of the t-butyl group that increased the energy gap by destabilizing the metal d$_\pi$ orbitals exerted by the π-donating bipyrazolate [85, 86].

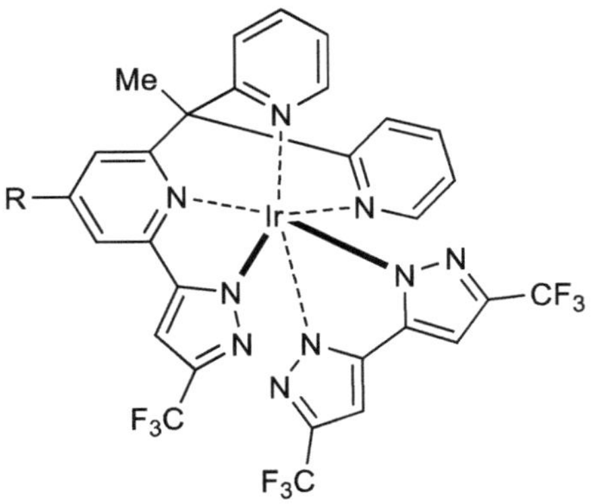

76   R = H, t-Bu

A molecular spring formed by hydrophobic MOF; Cu$_2$(tebpz) (tebpz = 3,3',5,5'-tetraethyl-4,4'-bipyrazolate (**77**)) and water as a nanoporous heterogeneous lyophobic system (HLS) was synthesized. This MOF was reported to have exceptional properties such as stability and efficiency at high operational pressure and temperatures [87].

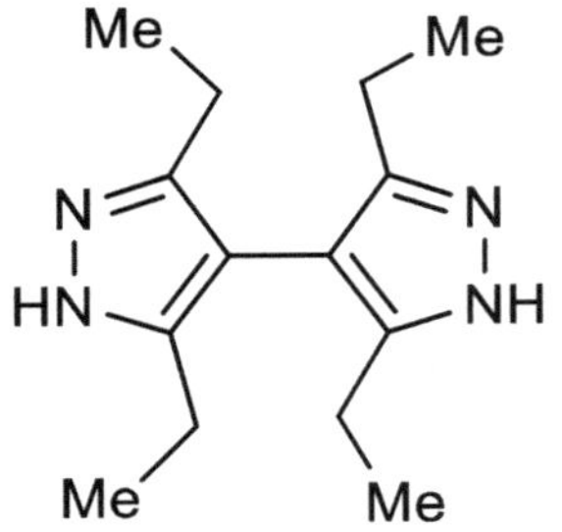

77

The solvothermal reaction of 3,3`,5,5`-tetramethyl-4,4`-bipyrazole (TMBP) (**31**) with nickel(II) chloride hexahydrate furnished a wavy layered nickel-based MOF material (Ni–TMBP). The obtained MOF was employed in the synthesis of lithium-ion battery anodes and demonstrated excellent specific capacity and cycling performance for efficient and stable electrochemical storage. Structure of the Ni–TMBP frameworks were established by single X-ray diffraction as stacking wavy layers with substantial flexibilities where each layer has a certain degree of structural changes during the electrochemical process of lithiation and delithiation. In addition, the abundant pores inside the Ni–TMBP frameworks provided enough pathways for Li$^+$ intercalation. The obtained results opened a new avenue for developing flexible layered MOFs for efficient and stable electrochemical storage [88].

Two Ag(I) coordination polymers based on bipyrazole and different dicarboxylates; [Ag(TMBP)(Hchda)]$_n$ (**78**) and [Ag$_2$(TMBP)$_2$(oba)]$_n$ (**79**) (H$_2$chda = *trans*-cyclohexane-dicarboxylic acid (**73**) and H$_2$oba = 4,4'-oxy-bis-benzoic acid (**80**)) were synthesized and characterized. The two compounds **78** and **79** showed interesting coordination features with 2-D and 3-D supramolecular networks. Compound **78** showed modest thermal stability and interesting solid-state fluorescent emission [89].

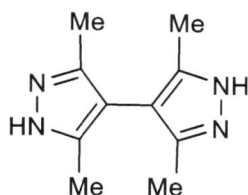

31 (TMBP)

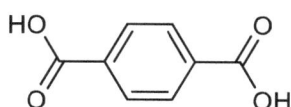

73 (H$_2$oba)

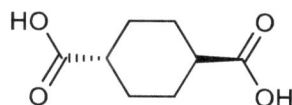

80 (H$_2$chda)

Six porous metal–organic frameworks (MOFs); {[Ni(BTC)$_{0.66}$(TMBP)$_2$] ·2MeOH·4H$_2$O}$_n$(**81**),{[Co(BTC)$_{0.66}$(TMBP)$_2$]·2MeOH·4H$_2$O}$_n$(**82**),{[Mn(BTC)$_{0.66}$ (TMBP)$_2$]·2MeOH·4H$_2$O}$_n$ (**83**), {[Cd(BDC)(TMBP)(H$_2$O)] ·2MeOH·DMF}$_n$ (**84**), {[Cd$_2$(NH$_2$-BDC)$_2$(TMBP)(H$_2$O)] ·MeOH·H$_2$O·DMF}$_n$ (**85**), and {[Co (BDC)(TMBP)(H$_2$O)]}$_n$ (**86**) (where H$_3$BTC = 1,3,5-benzenetricarboxylic acid (**59**)), H$_2$BDC = 1,4-benzenedicarboxylic acid (**57**), NH$_2$-H$_2$BDC = 2-amino-1-4-benzenedicarboxylic acid (**87**)), were synthesized and characterized. Complexes **81-85** exhibited selective CO$_2$ sorption over N$_2$ and CH$_4$. Complexes **81-83** also exhibited high water sorption at room temperature. Compounds **84** and **85** showed solid state emission properties with λ_{max} at 430, and 472 nm [90].

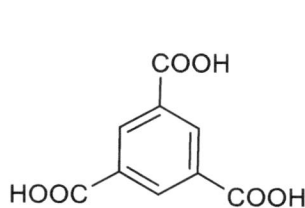

59 (H$_3$BTC)

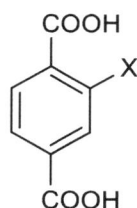

57, X = H (H$_2$BDC)
87, X = NH$_2$ (NH$_2$-H$_2$BDC)

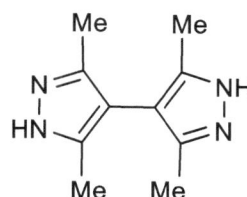

31 (TMBP)

The Ag(I) coordination polymer: [Ag$_2$(TMBP)$_3$(fum)]$_n$ (**88**, H$_2$fum = fumaric acid (**89**)), was prepared from 3,3',5,5'-tetramethyl-4,4'-bipyrazole (**31**) (TMBP) and maleic acid (**90**) (Scheme **9**) under hydrothermal conditions. Compound **88** exhibited photoluminescence in the solid state with an emission maximum at 470 nm upon excitation at 365 nm at room temperature. This was attributed to intra-ligand or/and inter-ligand $\pi \rightarrow \pi^*$ transition [91].

31

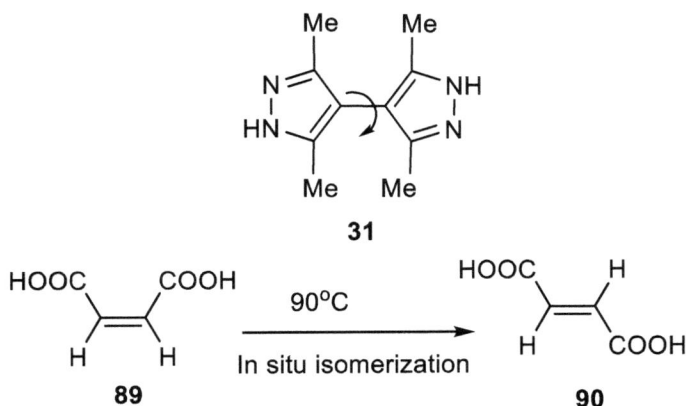

Scheme (9). The organic ligand is used in the construction of complex **88**.

Synthesis of seven Cu(II) coordination networks employing flexible 3,3′,5,5′-tetramethyl-4,4′-bipyrazole (TMBP) (**31**) and seven dicarboxylic acids **54, 91-94**, as auxiliary ligands under solvothermal conditions was reported. They demonstrated a number of interesting characters, such as supramolecular interactions, different dimensional frameworks and network topologies [92].

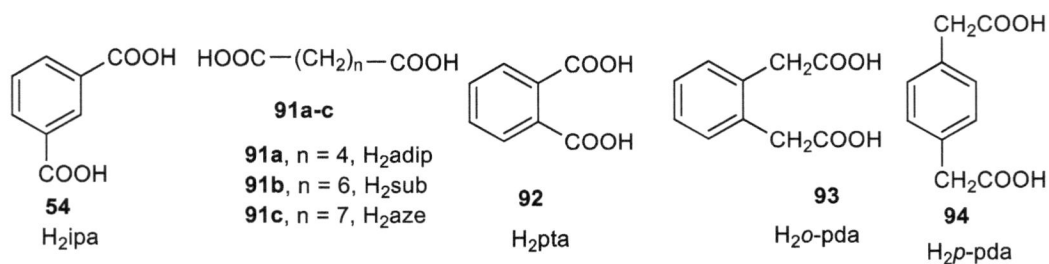

54
H₂ipa

91a-c

91a, n = 4, H₂adip
91b, n = 6, H₂sub
91c, n = 7, H₂aze

92
H₂pta

93
H₂o-pda

94
H₂p-pda

Encapsulation of tris(8-hydroxyquinolinato)aluminium (AlQ₃), a classic fluorescent molecule insensitive to oxygen, as a guest fluorophore into a highly porous coordination framework [Zn₄O(TMBP)₂(bdc)] (bdc = 1,4-benzenedicarboxylic acid (**57**)) resulted in a highly luminescent host–guest hybrid material AlQ₃@MAF-X10. Photoluminescence studies showed that AlQ₃@MAF-X10 exhibited yellowish green fluorescence [93].

A porous polymer framework **96** composed of a triazine unit bonded to three 3,5-dimethyl-bipyrazole moieties were synthesized from triazine **95** and 3,3′,5,5′-tetramethyl-4,4′-bipyrazole (**31**) (Scheme **10**). The product was used as a distinct building unit with strong fluorescence and porosity properties as well as for uptaking various types of metal species. When AgNO₃ was loaded, the solid framework demonstrated a brown color in response to water solutions of H_2S, even at dilution to 0.17 ppm, and a white-light emission was produced when an Ir(III) complex was doped (0.02% per weight) onto the framework [94].

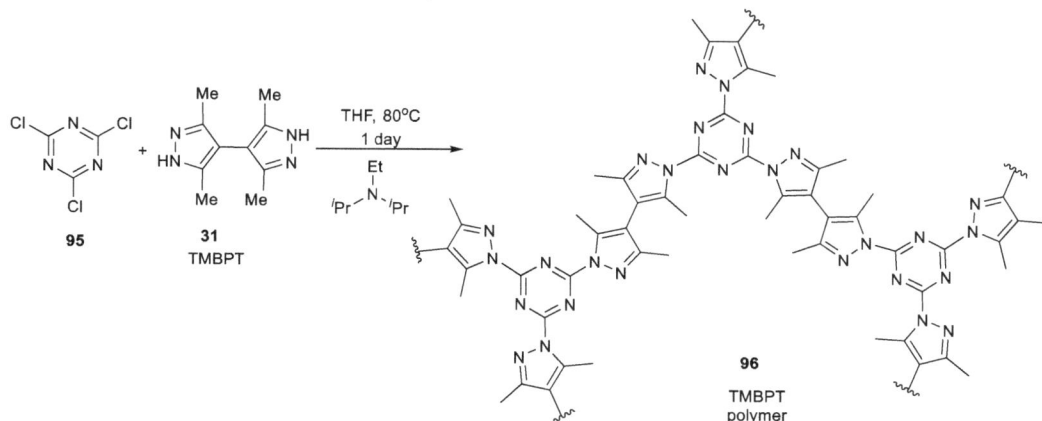

Scheme (10). Synthesis of the porous polymer framework **96**.

It was also reported that co-crystallization of 3,3′,5,5′-tetramethyl-4,4′-bipyrazole (**31**) (TMBP), with two aliphatic dicarboxylic acids, namely; suberic acid (**91b**) (H_2sub) and sebacic acid (**91c**) (H_2seb) resulted in the formation of two supramolecular solids [(TMBP)$_2$·(H_2sub)] (**97**) and [(TMBP)$_2$·(H_2seb)] (**98**). Single-crystal analyses described that the supramolecular solids **97** and **98** were 4-fold and 5-fold 2D → 2D parallel interpenetrated 6^3-hcb networks, respectively. The differences in the degree of interpenetration were referred to the different lengths of dicarboxylic acids. The photoluminescence investigation suggested that co-crystallization was employed to modulate the emission of the single component [95].

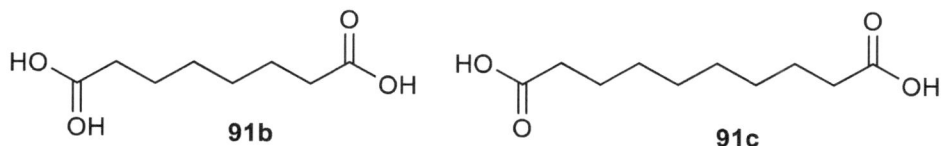

CONCLUSION

Bipyrazole derivatives were involved in the construction of a huge number of three-dimensional metal coordination compounds known as metal–organic frameworks (MOFs) with wide applications in different areas. The reported MOF were characterized by their high degree of crystallinity and internal porosity and were useful as photo-luminescence, sensing, gas separations, electrical conductivity and energy storage materials. The presented bipyrazoles with such high merits open new avenues for more exploration of their applications in different academic and industrial fields.

REFERENCES

[1]   Pettinari, C.; Pettinari, R.; Xhaferai, N.; Giambastiani, G.; Rossin, A.; Bonfili, L.; Eleuteri, A.M. Cuccioloni, M. Binuclear 3,3′,5,5′-tetramethyl-1H, H-4,4′-bipyrazole Ruthenium (II) complexes: Synthesis, characterization and biological studies. *Inorg. Chim. Acta,* **2020**, *513*, 119902.
[http://dx.doi.org/10.1016/j.ica.2020.119902]

[2]   Kanthecha, D.N.; Bhatt, B.S.; Patel, M.N.; Raval, D.B.; Thakkar, V.R.; Vaidya, F.U.; Pathak, C. Bipyrazole Based Novel Bimetallic μ-oxo Bridged Au (III) Complexes as Potent DNA Interacalative, Genotoxic, Anticancer, Antibacterial and Cytotoxic Agents. *J. Inorg. Organomet. Polym. Mater.,* **2020**, *30*(12), 5085-5099.
[http://dx.doi.org/10.1007/s10904-020-01618-2]

[3]   Zhang, Q.; Li, H.; Shao, Y.; Wang, Y.J. A New Cu (I)-Based Coordination Polymer: Crystal Structure, Molecular Docking and Protective Effect in Streptococcus-pneumoniae-Infected Mice by Promoting Immune Cell Response. *ChemistrySelect,* **2019**, *4*(37), 11019-11023.
[http://dx.doi.org/10.1002/slct.201901967]

[4]   Chi, Y.; Yeh, H.H. Luminescent platinum(II) complexes with biazolate chelates. *U.S. Patent,* **2015**, US 9040702 B1.

[5]   Hsu, C.W.; Ly, K.T.; Lee, W.K.; Wu, C.C.; Wu, L.C.; Lee, J.J.; Lin, T.C.; Liu, S.H.; Chou, P.T.; Lee, G.H.; Chi, Y. Triboluminescence and metal phosphor for organic light-emitting diodes: functional Pt (II) complexes with both 2-pyridylimidazol-2-ylidene and bipyrazolate chelates. *ACS Appl. Mater. Interfaces,* **2016**, *8*(49), 33888-33898.
[http://dx.doi.org/10.1021/acsami.6b12707] [PMID: 27960361]

[6]   Huang, M.J.; Deng, X.; Xian, W.R.; Liao, W.M.; He, J. Anion-directed structures and luminescences of two Cu (I) coordination polymers based on bipyrazole. *Inorg. Chem. Commun.,* **2019**, *101*, 121-124.
[http://dx.doi.org/10.1016/j.inoche.2019.01.025]

[7]   Liao, J.L.; Chi, Y.; Yeh, C.C.; Kao, H.C.; Chang, C.H.; Fox, M.A.; Low, P.J.; Lee, G.H. Near infrared-emitting tris-bidentate Os(II) phosphors: control of excited state characteristics and fabrication of OLEDs. *J. Mater. Chem. C Mater. Opt. Electron. Devices,* **2015**, *3*(19), 4910-4920.
[http://dx.doi.org/10.1039/C5TC00204D]

[8]   Pettinari, C.; Tăbăcaru, A.; Galli, S. Coordination polymers and metal–organic frameworks based on poly (pyrazole)-containing ligands. *Coord. Chem. Rev.,* **2016**, *307*, 1-31.
[http://dx.doi.org/10.1016/j.ccr.2015.08.005]

[9]   Mogensen, S.B.; Taylor, M.K.; Lee, J.W. Homocoupling Reactions of Azoles and Their Applications in Coordination Chemistry. *Molecules,* **2020**, *25*(24), 5950.
[http://dx.doi.org/10.3390/molecules25245950] [PMID: 33334079]

[10]   Mikus, M.S.; Sanchez, C.; Fridrich, C.; Larrow, J.F. Palladium Catalyzed C-O Coupling of Amino Alcohols for the Synthesis of Aryl Ethers. *Adv. Synth. Catal.,* **2020**, *362*(2), 430-436.
[http://dx.doi.org/10.1002/adsc.201901302]

[11]   Lavery, C.B.; Rotta-Loria, N.L.; McDonald, R.; Stradiotto, M. Pd$_2$dba$_3$/bippyphos: A robust catalyst system for the hydroxylation of aryl halides with broad substrate scope. *Adv. Synth. Catal.,* **2013**, *355*(5), 981-987.
[http://dx.doi.org/10.1002/adsc.201300088]

[12]   Gowrisankar, S.; Sergeev, A.G.; Anbarasan, P.; Spannenberg, A.; Neumann, H.; Beller, M. A general and efficient catalyst for palladium-catalyzed C-O coupling reactions of aryl halides with primary alcohols. *J. Am. Chem. Soc.,* **2010**, *132*(33), 11592-11598.

[http://dx.doi.org/10.1021/ja103248d] [PMID: 20672810]

[13]   Beaudoin, D.; Wuest, J.D. Synthesis of N-arylhydroxylamines by Pd-catalyzed coupling. *Tetrahedron Lett.,* **2011**, *52*(17), 2221-2223.
[http://dx.doi.org/10.1016/j.tetlet.2010.12.034]

[14]   Al-Fulaij, O.A.; Elassar, A.Z.A.; Dawood, K.M. Synthesis and characterization of new 3,3`-bipyrazole-4,4`-dicarboxylic acid derivatives and some of their palladium (II) complexes as pre-catalyst for Suzuki coupling reaction in water. *Eur. J. Chem.,* **2019**, *10*(4), 367-375.
[http://dx.doi.org/10.5155/eurjchem.10.4.367-375.1915]

[15]   Kotecki, B.J.; Fernando, D.P.; Haight, A.R.; Lukin, K.A. A general method for the synthesis of unsymmetrically substituted ureas *via* palladium-catalyzed amidation. *Org. Lett.,* **2009**, *11*(4), 947-950.
[http://dx.doi.org/10.1021/ol802931m] [PMID: 19178160]

[16]   Porzelle, A.; Woodrow, M.D.; Tomkinson, N.C. Palladium-catalyzed coupling of hydroxylamines with aryl bromides, chlorides, and iodides. *Org. Lett.,* **2009**, *11*(1), 233-236.
[http://dx.doi.org/10.1021/ol8025022] [PMID: 19035839]

[17]   Yu, S.; Haight, A.; Kotecki, B.; Wang, L.; Lukin, K.; Hill, D.R. Synthesis of a TRPV1 receptor antagonist. *J. Org. Chem.,* **2009**, *74*(24), 9539-9542.
[http://dx.doi.org/10.1021/jo901943s] [PMID: 19928811]

[18]   Ramdani, A.; Tarrago, G. Polypyrazolic macrocycles-I: A study of the polycondensation of 3-chloromethyl-3‘(5’),5-dimethyl-5′(3)-pyrazolyl-1-pyrazole. *Tetrahedron,* **1981**, *37*, 987-990.
[http://dx.doi.org/10.1016/S0040-4020(01)97674-4]

[19]   Radi, S.; Attayibat, A.; El-Massaoudi, M.; Bacquet, M.; Jodeh, S.; Warad, I.; Al-Showiman, S.S.; Mabkhot, Y.N.C. C,N-bipyrazole receptor grafted onto a porous silica surface as a novel adsorbent based polymer hybrid. *Talanta,* **2015**, *143*, 1-6.
[http://dx.doi.org/10.1016/j.talanta.2015.04.060] [PMID: 26078121]

[20]   Murakami, Y.; Yamamoto, T. Ni-Promoted Syntheses of New 3,3′-Dichloro-5,5′- bipyrazoles and Poly(bipyrazole-5,5′-diyl)s and Isolation of Nickel Complexes Relevant to the Syntheses. *Bull. Chem. Soc. Jpn.,* **1999**, *72*(7), 1629-1635.
[http://dx.doi.org/10.1246/bcsj.72.1629]

[21]   Schulze, M.C.; Scottb, B.L.; Chavez, D.E. A high density pyrazolo-triazine explosive (PTX). *J. Mater. Chem. A Mater. Energy Sustain.,* **2015**, *3*(35), 17963-17965.
[http://dx.doi.org/10.1039/C5TA05291B]

[22]   Tang, Y.; He, C.; Imler, G.H.; Parrish, D.A.; Shreeve, J.M. A C-C bonded 5,6-fused bicyclic energetic molecule: exploring an advanced energetic compound with improved performance. *Chem. Commun. (Camb.),* **2018**, *54*(75), 10566-10569.
[http://dx.doi.org/10.1039/C8CC05987J] [PMID: 30168821]

[23]   Kumar, D.; Tang, Y.; He, C.; Imler, G.H.; Parrish, D.A.; Shreeve, J.M. Multipurpose Energetic Materials by Shuffling Nitro Groups on a 3,3′-Bipyrazole Moiety. *Chemistry,* **2018**, *24*(65), 17220-17224.
[http://dx.doi.org/10.1002/chem.201804418] [PMID: 30231192]

[24]   Tang, Y.; Kumar, D.; Shreeve, J.M. Balancing excellent performance and high thermal stability in a dinitropyrazole fused 1,2,3,4-tetrazine. *J. Am. Chem. Soc.,* **2017**, *139*(39), 13684-13687.
[http://dx.doi.org/10.1021/jacs.7b08789] [PMID: 28910088]

[25]   Tang, Y.; He, C.; Imler, G.H.; Parrish, D.A.; Jean'ne, M.S. Energetic derivatives of 4,4′,5,5′-tetranitr-2H,2'H-3,3′-bipyrazole (TNBP): Synthesis, characterization and promising properties. *J. Mater. Chem. A Mater. Energy Sustain.,* **2018**, *6*(12), 5136-5142.
[http://dx.doi.org/10.1039/C7TA11172J]

[26]   Gospodinov, I.; Domasevitch, K.V.; Unger, C.C.; Klapötke, T.M.; Stierstorfer, J. Midway between

energetic molecular crystals and high-density energetic salts: Crystal engineering with hydrogen bonded chains of polynitro bipyrazoles. *Cryst. Growth Des.,* **2020**, *20*(2), 755-764.
[http://dx.doi.org/10.1021/acs.cgd.9b01177]

[27]   Domasevitch, K.V.; Gospodinov, I.; Krautscheid, H.; Klapötke, T.M.; Stierstorfer, J. Facile and selective polynitrations at the 4-pyrazolyl dual backbone: straightforward access to a series of high-density energetic materials. *New J. Chem.,* **2019**, *43*(3), 1305-1312.
[http://dx.doi.org/10.1039/C8NJ05266B]

[28]   Dalinger, I.L.; Suponitsky, K.Y.; Shkineva, T.K.; Lempert, D.B.; Sheremetev, A.B. Bipyrazole bearing ten nitro groups–a novel highly dense oxidizer for forward-looking rocket propulsions. *J. Mater. Chem. A Mater. Energy Sustain.,* **2018**, *6*(30), 14780-14786.
[http://dx.doi.org/10.1039/C8TA05179H]

[29]   Benabdellah, M.; Touzani, R.; Aouniti, A.; Dafali, A.; El Kadiri, S.; Hammouti, B.; Benkaddour, M. Inhibitive action of some bipyrazolic compounds on the corrosion of steel in 1 M HCl: Part I: Electrochemical study. *Mater. Chem. Phys.,* **2007**, *105*(2-3), 373-379.
[http://dx.doi.org/10.1016/j.matchemphys.2007.05.001]

[30]   Tebbji, K.; Oudda, H.; Hammouti, B.; Benkaddour, M.; Al-Deyab, S.S.; Aouniti, A.; Radi, S.; Ramdani, A. The effect of 1′,3,5,5′-tetramethyl-1′H-1,3′-bipyrazole on the corrosion of steel in 1.0 M hydrochloric acid. *Res. Chem. Intermed.,* **2011**, *37*(8), 985-1007.
[http://dx.doi.org/10.1007/s11164-011-0305-z]

[31]   Bouklah, M.; Hammouti, B.; Benkaddour, M.; Attayibat, A.; Radi, S. Corrosion inhibition of steel in hydrochloric acid solution by new bipyrazole derivatives. *Pigm. Resin Technol.,* **2005**, *34*(4), 197-202.
[http://dx.doi.org/10.1108/03699420510609088]

[32]   Abbas, A.A.; Abdellattif, M.H.; Dawood, K.M. Inhibitory activities of bipyrazoles: a patent review. *Expert Opin. Therap. Pat.,* **2021**. online ahead of print
[http://dx.doi.org/10.1080/13543776.2021.1953474]

[33]   Igarashi, T.; Sakurai, K.; Oi, T.; Obara, H.; Ohya, H.; Kamada, H. New sensitive agents for detecting singlet oxygen by electron spin resonance spectroscopy. *Free Radic. Biol. Med.,* **1999**, *26*(9-10), 1339-1345.
[http://dx.doi.org/10.1016/S0891-5849(98)00291-3] [PMID: 10381208]

[34]   Ohara, H.; Igarashi, T.; Sakurai, K.; Oi, T. Bipyrazole derivative, and medicine or reagent comprising the same as active component. **2000**, US 6121305A.

[35]   Igarashi, T.; Obara, H.; Oshii, T.; Sakurai, K. Bipyrazole derivative, and medicine and reagent consisting essentially thereof. *JP 10306077A,* **1998**.

[36]   Mayoral, E.P.; García-Amo, M.; López, P.; Soriano, E.; Cerdán, S.; Ballesteros, P. A novel series of complexones with bis- or biazole structure as mixed ligands of paramagnetic contrast agents for MRI. *Bioorg. Med. Chem.,* **2003**, *11*(24), 5555-5567.
[http://dx.doi.org/10.1016/j.bmc.2003.07.002] [PMID: 14642600]

[37]   Sarkar, A.; Mandal, T.K.; Rana, D.K.; Dhar, S.; Chall, S.; Bhattacharya, S.C. Tuning the photophysics of a bio-active molecular probe '3-pyrazolyl-2-pyrazoline'derivative in different solvents: dual effect of polarity and hydrogen bonding. *J. Lumin.,* **2010**, *130*(11), 2271-2276.
[http://dx.doi.org/10.1016/j.jlumin.2010.07.004]

[38]   Boldog, I.; Sieler, J.; Chernega, A.N.; Domasevitch, K.V. 4,4′-Bipyrazolyl: new bitopic connector for construction of coordination networks. *Inorg. Chim. Acta,* **2002**, *338*, 69-77.
[http://dx.doi.org/10.1016/S0020-1693(02)00902-7]

[39]   Ponomarova, V.V.; Komarchuk, V.V.; Boldog, I.; Chernega, A.N.; Sieler, J.; Domasevitch, K.V. Mixed-anion complexes with a bipyrazolyl ligand. A new entry to a realm of three-dimensional five-connected coordination topologies. *Chem. Commun. (Camb.),* **2002**, (5), 436-437.
[http://dx.doi.org/10.1039/b110599j] [PMID: 12120529]

[40] Boldog, I.; Rusanov, E.B.; Sieler, J.; Blaurock, S.; Domasevitch, K.V. Construction of extended networks with a trimeric pyrazole synthon. *Chem. Commun. (Camb.),* **2003**, (6), 740-741.
[http://dx.doi.org/10.1039/b212540d] [PMID: 12703800]

[41] Boldog, I.; Sieler, J.; Domasevitch, K.V. A unique polymeric coordination system that exhibits supramolecular isomerism within two dimensions. *Inorg. Chem. Commun.,* **2003**, *6*(6), 769-772.
[http://dx.doi.org/10.1016/S1387-7003(03)00101-1]

[42] Zhang, Z.X.; Huang, H.; Yu, S.Y. Synthesis and structure of a dipyrazol-bridged macrocyclic palladium (II) complex. *Wuji Huaxue Xuebao,* **2004**, *20*, 849-852.

[43] Tang, L.F.; Yang, P. Synthesis of dinuclear group 6 metal carbonyl complexes bridged by 4, 4'-bipyrazole ligands. *Trans. Met. Chem. (Weinh.),* **2004**, *29*(1), 31-34.
[http://dx.doi.org/10.1023/B:TMCH.0000014479.73155.44]

[44] Yu, S.Y.; Huang, H.P.; Li, S.H.; Jiao, Q.; Li, Y.Z.; Wu, B.; Sei, Y.; Yamaguchi, K.; Pan, Y.J.; Ma, H.W. Solution self-assembly, spontaneous deprotonation, and crystal structures of bipyrazolate-bridged metallomacrocycles with dimetal centers. *Inorg. Chem.,* **2005**, *44*(25), 9471-9488.
[http://dx.doi.org/10.1021/ic0509332] [PMID: 16323935]

[45] Zhang, G.F.; Lui, H.Q.; She, J.B.; Ng, S.W. Poly [[diaquabis (μ-3,3',5,5'-tetramethyl-4,4'-bipyrazole-κ²N:N') cobalt (II)] dinitrate]. *Acta Crystallogr.,* **2006**, *E62*(12), m3486-m3488.

[46] Boldog, I.; Rusanov, E.B.; Chernega, A.N.; Sieler, J.; Domasevitch, K.V. Coordination polymers of Co^{II} and 3,3',5,5'-tetramethyl-4,4'-bipyrazolyl: a novel metal-organic three-dimensional network with four-coordinated planar vertices. *J. Chem. Soc., Dalton Trans.,* **2001**, (6), 893-897.
[http://dx.doi.org/10.1039/b007183h]

[47] Domasevitch, K.V.; Boldog, I.; Rusanov, E.B.; Hunger, J.; Blaurock, S.; Schröder, M.; Sieler, J. Helical bipyrazole networks conditioned by hydrothermal crystallization. *Z. Anorg. Allg. Chem.,* **2005**, *631*(6-7), 1095-1100.
[http://dx.doi.org/10.1002/zaac.200400515]

[48] He, J.; Yin, Y.G.; Wu, T.; Li, D.; Huang, X.C. Design and solvothermal synthesis of luminescent copper(I)-pyrazolate coordination oligomer and polymer frameworks. *Chem. Commun. (Camb.),* **2006**, (27), 2845-2847.
[http://dx.doi.org/10.1039/b601009a] [PMID: 17007392]

[49] Zhang, J.P.; Horike, S.; Kitagawa, S. A flexible porous coordination polymer functionalized by unsaturated metal clusters. *Angew. Chem. Int. Ed.,* **2007**, *46*(6), 889-892.
[http://dx.doi.org/10.1002/anie.200603270] [PMID: 17183498]

[50] Zhang, J-P.; Kitagawa, S. 49Zhang, J.P.; Kitagawa, S. Supramolecular isomerism, framework flexibility, unsaturated metal center, and porous property of Ag (I)/Cu (I) 3,3`,5,5`-tetrametyl-4,-`-bipyrazolate. *J. Am. Chem. Soc.,* **2008**, *130*(3), 907-917.
[http://dx.doi.org/10.1021/ja075408b] [PMID: 18166049]

[51] Jozak, T.; Zabel, D.; Schubert, A.; Sun, Y.; Thiel, W.R. Ruthenium Complexes Bearing N–H Acidic Pyrazole Ligands. *Eur. J. Inorg. Chem.,* **2010**, *2010*(32), 5135-5145.
[http://dx.doi.org/10.1002/ejic.201000802]

[52] Mukherjee, S.; He, Y.; Franz, D.; Wang, S.Q.; Xian, W.R.; Bezrukov, A.A.; Space, B.; Xu, Z.; He, J.; Zaworotko, M.J. Halogen–C$_2$H$_2$ binding in ultramicroporous metal–organic frameworks (MOFs) for benchmark C$_2$H$_2$/CO$_2$ separation selectivity. *Chemistry,* **2020**, *26*(22), 4923-4929.
[http://dx.doi.org/10.1002/chem.202000008] [PMID: 31908047]

[53] Jiang, L.; Wang, J.; Gong, C.; Li, C.; Lu, L.; Li, H.; Singh, A.; Kumar, A.; Ma, A. Exploiting new 3D Cu (I)-based metal-organic framework as fluorescent sensor for nitroaromatics: An integrated experimental and computational investigation. *Inorg. Chem. Commun.,* **2019**, *106*, 18-21.
[http://dx.doi.org/10.1016/j.inoche.2019.05.022]

[54] El Ati, R.; Takfaoui, A.; El Kodadi, M.; Touzani, R.; Yousfi, E.B.; Almalki, F.A.; Hadda, T.B. Catechol oxidase and Copper (I/II) Complexes Derived from Bipyrazol Ligand: Synthesis, Molecular Structure Investigation of New Biomimetic Functional Model and Mechanistic Study. *Mater. Today Proc.*, **2019**, *13*, 1229-1237.
[http://dx.doi.org/10.1016/j.matpr.2019.04.092]

[55] Mouadili, A.; Attayibat, A.; El Kadiri, S.; Radi, S.; Touzani, R. Catecholase activity investigations using *in situ* copper complexes with pyrazole and pyridine based ligands. *Appl. Catal. A Gen.*, **2013**, *454*, 93-99.
[http://dx.doi.org/10.1016/j.apcata.2013.01.011]

[56] Grzywa, M.; Denysenko, D.; Schaller, A.; Kalytta-Mewes, A.; Volkmer, D. Flexible chiral pyrazolate-based metal–organic framework containing saddle-type $Cu^I_4(pyrazolate)_4$ units. *CrystEngComm*, **2016**, *18*(40), 7883-7893.
[http://dx.doi.org/10.1039/C6CE01594H]

[57] Wang, J.H.; Li, M.; Li, D. An exceptionally stable and water-resistant metal-organic framework with hydrophobic nanospaces for extracting aromatic pollutants from water. *Chemistry*, **2014**, *20*(38), 12004-12008.
[http://dx.doi.org/10.1002/chem.201403501] [PMID: 25081943]

[58] Grzywa, M.; Geßner, C.; Denysenko, D.; Bredenkötter, B.; Gschwind, F.; Fromm, K.M.; Nitek, W.; Klemm, E.; Volkmer, D. CFA-2 and CFA-3 (Coordination Framework Augsburg University-2 and-3); novel MOFs assembled from trinuclear Cu (I)/Ag (I) secondary building units and 3,3′,5,5′-tetraphenyl-bipyrazolate ligands. *Dalton Trans*, **2021**, *42*(19), 6909-6921.

[59] Vismara, R.; Tuci, G.; Mosca, N.; Domasevitch, K.V.; Di Nicola, C.; Pettinari, C.; Giambastiani, G.; Galli, S.; Rossin, A. Amino-decorated bis-(pyrazolate) metal–organic frameworks for carbon dioxide capture and green conversion into cyclic carbonates. *Inorg. Chem. Front.*, **2019**, *6*(2), 533-545.
[http://dx.doi.org/10.1039/C8QI00997J]

[60] Ling, X.Y.; Wang, J.; Gong, C.; Lu, L.; Singh, A.K.; Kumar, A.; Sakiyama, H.; Yang, Q.; Liu, J. Modular construction, magnetism and photocatalytic properties of two new metal-organic frameworks based on a semi-rigid tetracarboxylate ligand. *J. Solid State Chem.*, **2019**, *277*, 673-679.
[http://dx.doi.org/10.1016/j.jssc.2019.07.029]

[61] Jia, L.N.; Zhao, Y.; Hou, L.; Cui, L.; Wang, H.H.; Wang, Y.Y. An interpenetrated pillared-layer MOF: Synthesis, structure, sorption and magnetic properties. *J. Solid State Chem.*, **2014**, *210*(1), 251-255.
[http://dx.doi.org/10.1016/j.jssc.2013.11.008]

[62] Cai, S.L.; Lu, L.; Wu, W.P.; Wang, J.; Sun, Y.C.; Ma, A.Q.; Singh, A.; Kumar, A. A new mixed ligand based Cd (II) 2D coordination polymer with functional sites: photoluminescence and photocatalytic properties. *Inorg. Chim. Acta*, **2019**, *484*, 291-296.
[http://dx.doi.org/10.1016/j.ica.2018.09.066]

[63] Li, C.; Lu, L.; Wang, J.; Yang, Q.; Ma, D.; Alowais, A.; Alarifi, A.; Kumar, A.; Muddassir, M. Temperature tuned syntheses of two new d 10-based Cd (ii) cluster metal–organic frameworks: luminescence sensing and photocatalytic properties. *RSC Advances*, **2019**, *9*(51), 29864-29872.
[http://dx.doi.org/10.1039/C9RA05167H]

[64] Tomar, K.; Gupta, A.K.; Gupta, M. Change in synthetic strategy for MOF fabrication: From 2D non-porous to 3D porous architecture and its sorption and emission property studies. *New J. Chem.*, **2016**, *40*(3), 1953-1956.
[http://dx.doi.org/10.1039/C5NJ03044G]

[65] Sun, Y.Q.; Deng, S.; Ge, S.Z.; Liu, Q.; Chen, Y.P. A novel 2D dipyrazol-bridged cadmium (II) complex based on tetranuclear Cd_4O_4 clusters: synthesis, structure and luminescence. *J. Cluster Sci.*, **2013**, *24*(3), 605-617.
[http://dx.doi.org/10.1007/s10876-012-0527-2]

[66] Tomar, K. Assembly of a pcu topological porous Cd-trimesate framework with free carboxylic acid

group: Sorption and luminescent property. *Inorg. Chem. Commun.,* **2013**, *37*, 132-137.
[http://dx.doi.org/10.1016/j.inoche.2013.09.053]

[67]   Tomar, K. Assembly of an eight connected porous Cd (II) framework with octahedral and cubo-octahedral cages: Sorption and luminescent properties. *Inorg. Chem. Commun.,* **2013**, *37*, 127-131.
[http://dx.doi.org/10.1016/j.inoche.2013.09.065]

[68]   Zhang, E.; Jia, Q.; Zhang, J.; Ji, Z. Metal-Anion Coordination and Linker-Anion Hydrogen Bonding in the Construction of Metal-Organic Frameworks from Bipyrazole. *Chin. J. Chem.,* **2016**, *34*(2), 191-195.
[http://dx.doi.org/10.1002/cjoc.201500640]

[69]   Sun, Y.Q.; Deng, S.; Liu, Q.; Ge, S.Z.; Chen, Y.P. A green luminescent 1-D helical tubular dipyrazol-bridged cadmium(II) complex: a coordination tube included in a supramolecular tube. *Dalton Trans.,* **2013**, *42*(29), 10503-10509.
[http://dx.doi.org/10.1039/c3dt50620g] [PMID: 23752348]

[70]   Jin, S.; Huang, Y.; Fang, H.; Wang, T.; Ding, L. 3,5,3′,5′-Tetramethyl-4,4′-bi(1H-pyrazolyl) hemihydrate. *Acta Crystallogr.,* **2012**, *E68*(10), o2896.

[71]   Li, T.; Wang, F.L.; Wu, S.T.; Huang, X.H.; Chen, S.M.; Huang, C.C. A Highly Chemical and Thermal Stable Porous Metal–Organic Framework with Unusual 5-Connected zfy Topology. *Cryst. Growth Des.,* **2013**, *13*(8), 3271-3274.
[http://dx.doi.org/10.1021/cg400587j]

[72]   Dong, M.; Lu, L.; Tan, X.; An, B.; Singh, A.; Alowais, A.; Alarifi, A.; Kumar, A.; Muddassir, M. Syntheses and photocatalytic properties of two new d^{10}-and d^{9}-based 2D coordination polymers. *Inorg. Chim. Acta,* **2020**, *502*, 119334.
[http://dx.doi.org/10.1016/j.ica.2019.119334]

[73]   He, J.; Zhang, J.X.; Tan, G.P.; Yin, Y.G.; Zhang, D.; Hu, M.H. Second Ligand-Directed Assembly of Photoluminescent Zn(II) Coordination Frameworks. *Cryst. Growth Des.,* **2007**, *7*(8), 1508-1513.
[http://dx.doi.org/10.1021/cg070320v]

[74]   Wang, J.; Bai, C.; Hu, H.M.; Yuan, F.; Xue, G.L. A family of entangled coordination polymers constructed from a flexible V-shaped long bicarboxylic acid and auxiliary N-donor ligands: Luminescent sensing. *J. Solid State Chem.,* **2017**, *249*, 87-97.
[http://dx.doi.org/10.1016/j.jssc.2017.02.015]

[75]   Vismara, R.; Tuci, G.; Tombesi, A.; Domasevitch, K.V.; Di Nicola, C.; Giambastiani, G.; Chierotti, M.R.; Bordignon, S.; Gobetto, R.; Pettinari, C.; Rossin, A.; Galli, S. Tuning carbon dioxide adsorption affinity of zinc (II) MOFs by mixing bis (pyrazolate) ligands with N-containing tags. *ACS Appl. Mater. Interfaces,* **2019**, *11*(30), 26956-26969.
[http://dx.doi.org/10.1021/acsami.9b08015] [PMID: 31276365]

[76]   Liao, W.M.; Zeng, Q.; He, Y.; Duan, J.; He, J. Two homochiral crystals of anion-directed Cu (I) and Zn (II) helical coordination polymers. *J. Solid State Chem.,* **2019**, *277*, 448-453.
[http://dx.doi.org/10.1016/j.jssc.2019.07.003]

[77]   Baima, J.; Macchieraldo, R.; Pettinari, C.; Casassa, S. Ab initio investigation of the affinity of novel bipyrazolate-based MOFs towards H_2 and CO_2. *CrystEngComm,* **2015**, *17*(2), 448-455.
[http://dx.doi.org/10.1039/C4CE01989J]

[78]   Tabacaru, A.; Pettinari, C.; Timokhin, I.; Marchetti, F.; Carrasco-Marin, F.; Maldonado-Hódar, F.J.; Galli, S.; Masciocchi, N. Enlarging an isoreticular family: 3,3′,5,5′-tetramethyl-4,4′-bipyrazolato-based porous coordination polymers. *Cryst. Growth Des.,* **2013**, *13*(7), 3087-3097.
[http://dx.doi.org/10.1021/cg400495w]

[79]   Yu, F.; Li, J.; Cao, Z.H.; Kurmoo, M.; Zuo, J.L. Electrical conductivity of copper hexamers tuned by their ground-state valences. *Inorg. Chem.,* **2018**, *57*(6), 3443-3450.
[http://dx.doi.org/10.1021/acs.inorgchem.8b00243] [PMID: 29517912]

[80]   Mosca, N.; Vismara, R.; Fernandes, J.A.; Tuci, G.; Di Nicola, C.; Domasevitch, K.V.; Giacobbe, C.; Giambastiani, G.; Pettinari, C.; Aragones-Anglada, M.; Moghadam, P.Z.; Fairen-Jimenez, D.; Rossin, A.; Galli, S. Nitro-functionalized Bis(pyrazolate) Metal-Organic Frameworks as Carbon Dioxide Capture Materials under Ambient Conditions. *Chemistry,* **2018**, *24*(50), 13170-13180.
      [http://dx.doi.org/10.1002/chem.201802240] [PMID: 30028544]

[81]   Mosca, N.; Vismara, R.; Fernandes, J.A.; Casassa, S.; Domasevitch, K.V.; Bailón-García, E.; Maldonado-Hódar, F.J.; Pettinari, C.; Galli, S. CH₃-Tagged Bis (pyrazolato)-Based Coordination Polymers and Metal–Organic Frameworks: An Experimental and Theoretical Insight. *Cryst. Growth Des.,* **2017**, *17*(7), 3854-3867.
      [http://dx.doi.org/10.1021/acs.cgd.7b00471]

[82]   Li, G.P.; Zhang, K.; Zhao, H.Y.; Hou, L.; Wang, Y.Y. Increased Electric Conductivity upon I₂ Uptake and Gas Sorption in a Pillar-Layered Metal-Organic Framework. *ChemPlusChem,* **2017**, *82*(5), 716-720.
      [http://dx.doi.org/10.1002/cplu.201700063] [PMID: 31961526]

[83]   Tseng, C.H.; Fox, M.A.; Liao, J.L.; Ku, C.H.; Sie, Z.T.; Chang, C.H.; Wang, J.Y.; Chen, Z.N.; Lee, G.H.; Chi, Y. Luminescent Pt (II) complexes featuring imidazolylidene–pyridylidene and dianionic bipyrazolate: from fundamentals to OLED fabrications. *J. Mater. Chem. C Mater. Opt. Electron. Devices,* **2017**, *5*(6), 1420-1435.
      [http://dx.doi.org/10.1039/C6TC05154E]

[84]   Gajera, S.B.; Mehta, J.V.; Kanthecha, D.N.; Patel, R.R.; Patel, M.N. Novel cytotoxic oxovanadium (IV) complexes: Influence of pyrazole-incorporated heterocyclic scaffolds on their biological response. *Appl. Organomet. Chem.,* **2017**, *31*(11), e3767.
      [http://dx.doi.org/10.1002/aoc.3767]

[85]   Li, Y.S.; Liao, J.L.; Lin, K.T.; Hung, W.Y.; Liu, S.H.; Lee, G.H.; Chou, P.T.; Chi, Y. Sky blue-emitting iridium (III) complexes bearing nonplanar tetradentate chromophore and bidentate ancillary. *Inorg. Chem.,* **2017**, *56*(16), 10054-10060.
      [http://dx.doi.org/10.1021/acs.inorgchem.7b01583] [PMID: 28796502]

[86]   Liao, J.L.; Chi, Y.; Sie, Z.T.; Ku, C.H.; Chang, C.H.; Fox, M.A.; Low, P.J.; Tseng, M.R.; Lee, G.H. Ir (III)-Based phosphors with bipyrazolate ancillaries; rational design, photophysics, and applications in organic light-emitting diodes. *Inorg. Chem.,* **2015**, *54*(22), 10811-10821.
      [http://dx.doi.org/10.1021/acs.inorgchem.5b01835] [PMID: 26529058]

[87]   Grosu, Y.; Li, M.; Peng, Y.L.; Luo, D.; Li, D.; Faik, A.; Nedelec, J.M.; Grolier, J.P. A highly stable nonhysteretic {Cu₂ (tebpz) MOF + water} molecular spring. *ChemPhysChem,* **2016**, *17*(21), 3359-3364.
      [http://dx.doi.org/10.1002/cphc.201600567] [PMID: 27442186]

[88]   An, T.; Wang, Y.; Tang, J.; Wang, Y.; Zhang, L.; Zheng, G. A flexible ligand-based wavy layered metal-organic framework for lithium-ion storage. *J. Colloid Interface Sci.,* **2015**, *445*, 320-325.
      [http://dx.doi.org/10.1016/j.jcis.2015.01.012] [PMID: 25638743]

[89]   Huang, X.H.; Xiao, Z.P.; Wang, F.L.; Li, T.; Wen, M.; Wu, S.T. Syntheses and characterizations of two silver (I) coordination polymers constructed from bipyrazole and dicarboxylate ligands. *J. Coord. Chem.,* **2015**, *68*(10), 1743-1753.
      [http://dx.doi.org/10.1080/00958972.2015.1022165]

[90]   Tomar, K.; Rajak, R.; Sanda, S.; Konar, S. Synthesis and Characterization of Polyhedral-Based Metal–Organic Frameworks Using a Flexible Bipyrazole Ligand: Topological Analysis and Sorption Property Studies. *Cryst. Growth Des.,* **2015**, *15*(6), 2732-2741.
      [http://dx.doi.org/10.1021/acs.cgd.5b00056]

[91]   Han, L.L.; Wang, Y.X.; Guo, Z.M.; Yin, C.; Hu, T.P.; Wang, X.P.; Sun, D. Synthesis, crystal structure, thermal stability, and photoluminescence of a 3-D silver (I) network with twofold interpenetrated dia-f topology. *J. Coord. Chem.,* **2015**, *68*(10), 1754-1764.

[http://dx.doi.org/10.1080/00958972.2015.1028380]

[92]   Han, L.L.; Wang, S.N.; Jagličić, Z.; Zeng, S.Y.; Zheng, J.; Li, Z.H.; Chen, J.S.; Sun, D. Synthesis, structural versatility and magnetic properties of a series of copper (II) coordination polymers based on bipyrazole and various dicarboxylate ligands. *CrystEngComm,* **2015**, *17*(6), 1405-1415.
[http://dx.doi.org/10.1039/C4CE02248C]

[93]   Lin, R.B.; Zhou, H.L.; He, C.T.; Zhang, J.P.; Chen, X.M. Tuning oxygen-sensing behaviour of a porous coordination framework by a guest fluorophore. *Inorg. Chem. Front.,* **2015**, *2*(12), 1085-1090.
[http://dx.doi.org/10.1039/C5QI00157A]

[94]   Liu, J.; Yee, K.K.; Lo, K.K.W.; Zhang, K.Y.; To, W.P.; Che, C.M.; Xu, Z. Selective Ag(I) binding, H$_2$S sensing, and white-light emission from an easy-to-make porous conjugated polymer. *J. Am. Chem. Soc.,* **2014**, *136*(7), 2818-2824.
[http://dx.doi.org/10.1021/ja411067a] [PMID: 24456260]

[95]   Han, L.L.; Li, Z.H.; Chen, J.S.; Wang, X.P.; Sun, D. Solution and mechanochemical syntheses of two novel cocrystals: ligand length modulated interpenetration of hydrogen-bonded 2D 63-hcb networks based on a robust trimeric heterosynthon. *Cryst. Growth Des.,* **2014**, *14*(3), 1221-1226.
[http://dx.doi.org/10.1021/cg4017454]

SUBJECT INDEX

A

B

C

www.ingramcontent.com/pod-product-compliance
Lightning Source LLC
Chambersburg PA
CBHW050519240326
41598CB00086B/606